BRITAIN'S JET AGE

From the Meteor to the Sea Vixen

Guy Ellis

AMBERLEY

Front cover: Meteor T.7 WA591/G-BWMF. Built in 1949, it was restored to flying condition by Meteor Flight and then Air Atlantique's Classic Flight. Seen here in her element. (Classic Air Force)

Back cover: Gloster E28/39 replica, Gloster Meteor F8 WH3644 and Gloster Javelin FAW9 XH90 represent Gloster Jets from the first to the last. Photographed at Jet Age Museum. (Richard Trinder)

First published 2016

Amberley Publishing
The Hill, Stroud
Gloucestershire, GL5 4EP

www.amberley-books.com

ISBN 978 1 4456 4900 9 (print)
ISBN 978 1 4456 4901 6 (ebook)

British Library Cataloguing in Publication Data.
A catalogue record for this book is available from the British Library.

Typeset in 9.5pt on 12pt Celeste.
Typesetting by Amberley Publishing.
Printed in the UK.

Only the Beginning

The mercurial British weather had delayed until early evening what was to be an auspicious event on 15 May 1941. As patches of blue sky at last began to appear at 7.40 p.m., the ground crew hurriedly prepared the Gloster E28/39. Test pilot Gerry Sayer ran through his pre-flight checks: flaps retracted, elevator trim set and sliding hood locked open. He then taxied to the eastern end of RAF Cranwell's long runway, pushed the throttle forward against the brakes, and as he released them the aircraft accelerated before gently lifting into the air within 600 yards.

Sayer climbed to 1,000 feet, retracted the undercarriage and disappeared into the clouds. Those on the ground were the first to hear the distinct sound of an unseen jet, something we now take for granted. The E28 reappeared as the pilot made a wide circuit of the airfield, preparing his approach and landing. After a trouble-free flight lasting 17 minutes, he touched down on the main wheels, the nose dropping gently onto the forward undercarriage.

It may be said that this flight represents one of the most important moments in history – changing our lives forever, revolutionising warfare and travel and creating new products and technology. It was at Cranwell that Frank Whittle had developed his idea to overcome the limitations of propeller propulsion by using a jet engine, and it was his Whittle W1 jet that had powered the E28/39.

Although the German Heinkel He 178 was the first jet to fly, it was from the Whittle jet engine – patented in 1930 – that all jet aircraft are descended. Wartime secrecy and the potential to develop a weapon that could vanquish the Nazis meant this momentous event went unannounced and virtually unrecorded.

> It's only the beginning but the implications are terrific.
>
> Gerald Sayer, 1941

Sir Frank Whittle. (Jet Age Museum)

Frank Whittle congratulates Gerry Sayer after the first flight of the E28/39. (Jet Age Museum)

The E28/39 replica on display at the Jet Age Museum, Gloucestershire. (Richard Trinder)

Sir Isaac Newton's Third Law of Motion

Fun at any age is to inflate a balloon and then release it to make it perform a crazy uncontrolled flight. As we all know, it moves in the opposite direction to the outflow of air, which is an example of thrust in its rawest form.

A jet engine operates on a similar principle. Air flows through the engine via the intake at the front, is compressed and then forced into combustion chambers, where fuel is injected and the resulting mixture ignited. The gases formed expand and are then released from the rear of the combustion chamber. This backward movement of heated, expanded air exerts an equal and opposite force propelling the aircraft forward – known as 'thrust'. As the gases are expelled, they pass through the turbine, a set of fan-like blades rotating a shaft that, in its turn, rotates the compressor, which, again, brings in a new supply of air.

The first gas-turbine engine was patented and built by John Barber of Nottingham in 1792. This was a chain-driven engine with a compressor, combustion chamber and turbine, powered by wood, coal and other inflammable substances. However, the heat-resistant materials necessary to generate enough power to compress the air and gases, and to have sufficient power to effectively drive the chain and any attached devices, were simply not available. These same limitations meant that Frenchman Maxime Guillaume was unable to build his multi-stage axial-flow turbojet, patented in May 1921. In a 1926 paper, *An Aerodynamic Theory of Turbine Design*, Alan Arnold Griffith laid out the founding principles of a turboprop engine using an axial compressor fitted with aerofoil shaped blades, creating more efficient airflow.

By that same year, Frank Whittle had developed his theory of using gas turbines to power aircraft at speeds and altitudes that could not be achieved by piston engines. He patented his idea for a centrifugal compressor-driven turbojet engine in 1930.

An axial-flow engine compresses air through a set of fans on the same axis. With a centrifugal compressor the air is forced outwards into compression chambers, changing the direction of the airflow. These compressors are more robust and easier to manufacture, while the axial-flow compressor provides higher pressure ratios and therefore more thrust.

Whittle's patent and papers were carefully studied in Germany by Dr Hans von Ohain, who registered his patent for a turbojet in 1936. Working for Ernst Heinkel, in a secret private venture, he developed a jet engine which flew for the first time on 27 August 1939 in an He 178. Captain Erich Warstiz made the first jet aircraft flight, which was marked by the loss of the engine shortly after take-off when a bird was sucked into the intake – the world's first jet–bird strike.

Largely ignored by the Reichsluftfahrtministerium, the development of the jet was hampered by the demand to produce 'normal' aircraft. A less-developed ducted-fan jet system was used to power the Italian-built Caproni-Campini CC2, which became the world's second jet aircraft to fly, in 1937.

Apart from the turbojet, both axial and centrifugal, there are four other main types of jet engine. The first is the turboprop. Here, exhaust gases drive a propeller, which provides improved efficiency at speeds below 500 mph, typically fitted to light aircraft and smaller short-haul passenger aircraft. With development, higher speeds have been achieved; consequently, propellers are smaller in diameter, with multiple scimitar-shaped blades that increase the airflow into the compressor.

Secondly, most modern airliners are equipped with turbofan engines where a very large fan at the front sucks in air, but most of that air flows around the engine rather than through the compressor, pushing additional air out of the rear of the engine without increasing fuel consumption. This has the added advantage of quieter engines.

Thirdly, for helicopters specifically, the turboshaft engine was developed, which drives the rotor blades at a different speed to that of the compressor. As the rotor blade provides lift and directional control, it is important that it maintains a constant speed even while extra power is provided by the engine.

Fourthly, the ramjet is a simple engine with no moving parts. Here forward speed rams air into the front of the engine and so requires an assisted take-off, such as an air launch or a catapult. Thrust is then dependent on increasing the forward speed. These engines are used in high-speed, short-burst environments such as missiles.

Jet engines are lighter, have fewer moving parts, provide greater power and are more fuel efficient than piston engines. The Supermarine Spitfire had on average a top speed of 367 mph; the first British jet, the Gloster E28/39, produced 850 lbs of forward thrust and a top speed of 466 mph, whereas the Concorde could produce 38,050 lbs of thrust per engine with a top speed of 1,334 mph.

> To every action there is an equal and opposite reaction.
>
> Third Law of Motion, Sir Issac Newton (1642–1727)

The He 178 original was destroyed in 1943, but this replica is on display in the arrivals hall of Rostock-Laage airport. (Gryffindor – Creative Commons)

Photographed in 1940, the Campini Caproni CC2 had a fuselage-mounted conventional liquid-cooled piston engine, driving a turbine which drove the compressor. (Phil DeFer)

A Campini-Caprini CC2, either MM487 or MM488, on display near Rome. (Alan Wilson)

Above left: The cockpit of the replica E28/39. (Jet Age Museum)

Above right: An original, incomplete Welland centrifugal engine on display at Jet Age Museum. (Guy Ellis)

The Rolls-Royce Derwent Mk 8 centrifugal turbojet provided 3,600 lbf (16.0 kN) of thrust and was used in the Meteor. (Guy Ellis)

The Armstrong Siddeley Sapphire axial turbojet engine provided 8,300 lbf (37,000 kN), and was used on the Gloster Javelin FAW Mk 1. (Guy Ellis)

Rolls-Royce Welland No. 183 is a working example owned by Sir James Dyson. (Guy Ellis)

The first operational jet fighter was the German Messerschmitt Me 262, powered by two wing-mounted Junkers Jumo turbojet units. (Amberley Publishing)

In 1945 the first turboprop aircraft was flown using a converted Meteor I, EE227, fitted with Rolls-Royce Trent engines. (Mike Dowsing)

Ignored, Rejected and Given Away

In his 1928 thesis, *Future Developments in Aircraft Design,* written while at Cranwell College, Whittle proposed that the only way to make aircraft fly faster and further was to fly them at higher altitudes, but at these heights propellers do not work efficiently in the thinner air.

While an apprentice in the Royal Air Force, Whittle's prodigious mathematical talent was picked up and in 1926 he was recommended for officer and pilot training. While at Central Flying School in 1929, he developed his turbojet and was granted a patent in April 1931 for 'a reaction motor suitable for aircraft propulsion'. He sent his work to the Air Ministry, where Alan Arnold Griffith found an error in Whittle's calculations and became convinced that the frontal air intake was impractical and that the exhaust thrust was insufficient to generate sufficient forward power.

Whittle proved to be an outstanding pilot and in 1931 he was appointed as a floatplane and catapult test pilot at the Marine Aircraft Experimental Establishment, Felixstowe. Flying time was limited and he continually worked on and submitted various ideas to improve bomb loading and anchor design and to expand the field of fire, even creating a prototype enclosed gun turret. His contribution to test flying saw him posted to the School of Engineering, Henlow, followed in 1934 by admission to Peterhouse College, Cambridge, where he was encouraged to continue his work on turbines.

By now married and with a young family, he did not have the £5 necessary to renew his patent. Rolf Williams, a fellow pilot from Felixstowe, contacted him in May 1935 and with his business partner, Collingwood Tinling, proposed covering the cost of the patent and raising the necessary funds to build the engine. With backing from investment bankers, Power Jets Ltd was formed in March 1936 but it was over a year before the first Whittle unit was run.

The threat of war encouraged government interest in the potential of the turbojet and in July 1939 the Air Ministry contracted Power Jets to build a jet engine to power an experimental aircraft to be built by Gloster Aircraft Company. The E28/39 flights proved the jet engine concept and it was vital that development was accelerated. Harold Roxbee Cox, Baron Kings Norton, created the Gas Turbine Collaboration Committee in 1941 as a forum where companies working on jet-engine projects could share designs and test results. The Rover Company was contracted to produce a production gas turbine but after a number of delays the contract was handed to Rolls-Royce under the supervision of Stanley Hooker. Baron Kings Norton shared the jet-engine technology with the Americans in July 1941 'and soon the unparalleled industrial might of the USA was being applied to the revolutionary concept'. (Reprinted with the kind permission of Cranfield University) Working with the Americans highlighted to Whittle how much more he could have achieved in the United Kingdom if he had had the same level of support from the British Air Ministry.

In 1944, the Minister for Aircraft Production, Sir Stafford Cripps, nationalised Power Jets and merged it into the Royal Aircraft Establishment to create the National Gas Turbine Establishment at Pyestock. Feeling unappreciated, Whittle and his team resigned.

Britain lost its lead in jet technology by giving it away. All the development data on supersonic flight were handed to the Americans, who went on to be the first to break the speed of sound and the post-war government gave twenty-five Rolls-Royce-developed Nene engines to the Russians.

The last straw for Frank Whittle was the cancellation of the Miles M52 supersonic project, and he ceased all work on jet-engine technology. He later moved to the United States, where his talent was both appreciated and funded.

> Whittle is one of the great engineers of the century, perhaps the greatest if judged by the scale of the consequences of his invention.
>
> From *A Wrack Behind* by Lord Kings Norton

Royal Air Force College Cranwell, Lincolnshire, UK. (Dagomir Oniwenko)

Above left: Sir Stanley Hooker (left), who overcame production problems with the Welland, and then designed the Derwent engine, was firm friends with Sir Frank Whittle (right). (Stephen Greenstead)

Above right: Sir Frank Whittle and Sir Stanley Hooker, seen inspecting an engine combustion chamber. (Stephen Greenstead)

The Lookheed XP-80 was built in 1943, around the blueprint dimensions of a British Halford H-1B turbojet. (Lockheed Martin)

The USSR reverse-engineered the Nene engines to produce the Klimov range of engines that powered the famous MiG-15.

The memorial to Frank Whittle near Lutterworth, where the engine for the Gloster E.28/39 was produced. (G-Man)

The War Years

Gloster Meteor

F/O 'Dixie' Dean lifted off from RAF Manston late in the afternoon of 4 August 1944. He was flying EE216, one of 616 Squadron's new Meteor Mk Is, code YQ-E. Spotting a target, Dean dived down from 4,500 feet, reaching a speed of 450 mph, lined up for an attack from the stern and pressed the fire button. The four 20 mm cannons jammed and nothing happened. Desperate to destroy the German flying bomb, he flew alongside and carefully manoeuvred his port wing tip under the stubby wing of the Diver. He pulled up sharply and tipped the enemy missile off balance, putting it into a dive which ended with an explosion 4 miles south of Tonbridge.

This was the first jet-to-jet combat in history, followed 20 minutes later when fellow squadron member F/O J. K. Rodger shot down a Diver near Tenterden.

Because the maximum thrust produced by Whittle's engine was insufficient to power a fighter, Gloster's designer George Carter opted for a twin-engine design. The Air Ministry ordered twelve prototypes in February 1941 and Gerry Sayer began testing the first aircraft in July 1942. This aircraft was fitted with non-flying, low-thrust engines for taxiing trials.

Problems developing the Whittle unit saw the prototypes fitted with an array of different engine types and configurations. After Gerry Sayer had been lost testing a Hawker Typhoon, Michael Daunt took over the flying programme, piloting the first Meteor to fly on 5 March 1943. This was the fifth prototype, DG206/G, which was powered by De Havilland turbojets. Prototype Meteor F9/40 DG202/G was the first Meteor to be powered by the Rover-built W2B Welland engine, flying on 24 July 1942. It was also used for carrier handling trials aboard HMS *Pretoria Castle* in August 1945.

Delivery of the production Meteor I took place a year later when 616 (South Yorkshire) Squadron, based at Culmhead, became the world's first operational jet squadron, predating the August combat debut of the Me 163 rocket-powered fighter, and the first Me 262 operational unit formed in the October. The first aircraft, EE215, was delivered on 12 July 1944, followed a day later by the soon to be famous EE216 and EE217. Everything was done to keep them secret but their unfamiliar whine and mystifying lack of propellers made this impossible.

Although no dual-seat trainers were available, conversion to the new aircraft proved easy: the pilots would taxi the aircraft for a short time and simply take off. Royal Canadian Air Force F/O William McKenzie recalled that:

> The most difficult thing was to get accustomed to jet flight. It was very quiet because you were up in front of the engines. All I could do was sit there looking at the holes where the props should

have been, and thinking, 'I see it, but I don't believe it! What's holding me up?' [*Legion Magazine*]

The squadron moved to RAF Manston to begin operations against the V1 flying bombs, but the Meteors proved to be underpowered, and the pilots complained about heavy controls and poor visibility to the sides and rear. New flying techniques had to be learned to maintain airspeed as there was a significant delay between increasing the throttle and the corresponding reaction from the engines. These reports caused doubts within the Air Ministry as to the wisdom of spending more time and money on jet propulsion, but those who saw the potential won the day.

Battle of the Divers

The first wave of the German V1 missile attacks on Britain began on 13 June 1944, when one struck a railway bridge near Bethnal Green, killing six people and injuring thirty. The pulse-jet-powered bombs were terrifying as there was no knowing where they would fall.

Known as 'Buzz bombs' or 'Doodle bugs', due to the distinctive sound made by the pulse jet, the RAF code-named them 'Divers'. Combating this menace was both important and difficult. The Divers were able to reach a speed of more than 400 mph, marginally faster than the best of the piston-engine fighters, even those that had been stripped of their armour and paint, were highly polished and powered by the highest octane fuel available.

A series of lines of defence were set up in the South East to counter the flying bomb menace. The first was radar that would detect the incoming missiles, the second were the heavy anti-aircraft batteries established on the North Downs, and the third line was made up of a balloon barrage set to protect the most vital or common targets.

The RAF's new jet fighters joined the battle towards the end of the summer offensive, but they were limited in number and restricted by their short range. The 410 mph Meteor was marginally faster than the Tempest V and Spitfire XIV with top speeds of 405 mph and 396 mph, respectively, but it excelled at 1,000 to 3,000 feet, exactly where the V1 operated.

The squadron claimed twelve and a half V1s destroyed, far below the success rate of the piston-engine fighters, but they were the pioneers of jet combat, proving the capability of the jet against high-speed targets.

During the second phase of the German rocket attacks, 616 Squadron began to receive the Meteor F Mk III. This was the first really capable RAF jet and introduced an improved bubble canopy, a greater fuel capacity and the new Rolls-Royce Derwent engines. Within a few short months dramatic developments in jet technology had increased the power of the F Mk III from 2,000 lbs to 3,500 lbs thrust per engine.

Europe

In January 1945, a single flight from 616 Squadron was despatched to Nijmegen in the Netherlands. Their aircraft, painted gloss white to prevent confusion with the Luftwaffe's Me 262, were restricted to operating over Allied-held territory only. All the pilots were keen to engage the German jet in combat, but this never happened; instead, they were tasked with armed reconnaissance and ground-attack operations. The remainder of 616 Squadron arrived in Europe during April after intensive training and further operations against the V1 missiles.

Confidence in the Meteor F Mk III and a reduction in the chances that one could be captured saw 504 (County of Nottingham) Squadron become the second jet unit sent to Europe in March 1945. Such was the secrecy surrounding jet fighters that it was only towards the end of February 1945 that the BBC announced that Meteors were undertaking operational flights.

De Havilland Vampire

The De Havilland Halford H1 turbojet, which gave rise to the legendary Goblin engine, first powered a Meteor prototype and then a Meteor Mk II. This did not go into production, because De Havilland reserved the Goblin for their Vampire fighter.

Before the Air Ministry had issued a specification for an experimental single-engine jet aircraft armed with four 20 mm cannon, De Havilland staff had examined the Gloster E28/39 and were handed all the development details. Unusually, the Air Ministry allowed De Havilland to develop both the engine and the aircraft, when normally each component had to be built by separate organisations. Sir Geoffrey de Havilland, designer Ronald Bishop and engine designer Frank Halford developed the 'Spider Crab', the code word for the soon-to-be-famous Vampire.

Following a rapid sixteen-month development programme, De Havilland's chief test pilot, Geoffrey de Havilland Jr, made the first flight on 20 September 1943. The 30-minute flight revealed a number of problems, the most serious being a tendency to snake. This peculiarity was attributed to the tall tail fins, which were quickly squared off.

Assigned the De Havilland company code DH100, the aircraft was – in the spring of 1944 – the first Allied aircraft to exceed 500 mph, a short time before a Lockheed Shooting Star using a Goblin unit that had been shipped to the United States.

The first production Vampire flew in April 1945, with the first deliveries to 247 Squadron in March 1946. Unlike other wartime aircraft, the order for 300 Vampires was not cancelled; rather, development continued, and engines with an increased thrust of 3,100 lbs, a pressurised cockpit and a bubble canopy were all introduced after the first fifty aircraft.

Meteor F Mk I EE216, in which F/O 'Dixie' Dean scored the first confirmed victory by a British jet fighter. (www.1000aircraftphotos.com)

Meteor III EE239 Q is the aircraft on the box artwork for the 1/72 Airfix kit. (Gerald Trevor Roberts, via John Roberts)

In the nose of a V1 was a rotor that had been turned a set number of times to bring the missile over the selected target; when the rotor stopped, the bomb would go into a steep dive and its 1-ton explosive warhead would detonate on impact. (PanzerDB)

The Royal Observer Corps backed up the radar defences as the technology was in its infancy and the operators often experienced difficulty in identifying a Diver. (Amberley Publishing)

Light anti-aircraft batteries and searchlight sites supported the heavy AA batteries situated behind them, which created a formidable barrier to the flying bombs. (Amberley Publishing)

A V1 was small, fast and difficult to see, while the favoured approach from astern only provided a 16-foot aiming point, which was obscured by the haze from the pulse jet; if an attacking pilot was too close, he was in danger from the bomb exploding. (PanzerDB)

Meteor YQ-B, possibly EE236, being refuelled at Lübeck, Germany, in June 1945. (Gerald Trevor Roberts, via John Roberts)

Gloster Meteor F Mk IIIs of 616 Squadron, led by S/L Dennis, lined up at Kastrup Airport ahead of the Copenhagen Air Pageant held on 1 July 1945 to raise funds for a bombed French school. (John Donkersley)

The show and high-speed, low-level pass by Squadron Leader Tony Gaze, watched by a crowd of 250,000, was described in the Danish aviation magazine *Flyv*. 'At first these fantastic machines showed their impressive speed in a perfect formation fly past. Then, one of the nameless "Meteor" pilots performed a spectacular aerobatic show. He rose to the sky at 1,000 kilometres per hour and then dived to the ground at a speed faster than that of the sound.' (John Donkersley)

Meteor EE243, YQ-N, shortly before 616 Squadron was disbanded on 29 August 1945 at Lubeck. (Gerald Trevor Roberts, via John Roberts)

This English Electric-built Vampire F1, TG302, is shown with its starter ready for action in 1947. (R. A. Scholefield)

Britain Leads the Way

Post-war, the British aviation industry focused on addressing metallurgical and technical problems, and exploring the aerodynamic notion of compressibility, popularly known as the 'sound barrier'. It was generally agreed that flight beyond the speed of sound was impossible.

'Mach' is the term used to indicate the speed of sound and if an aircraft is travelling at 80 per cent of Mach, this is expressed as Mach 0.8. When almost achieving Mach, the flow of air over the aircraft becomes disrupted – this, for instance, in the early Vampires, would cause a series of high-speed stalls.

The heroes of the day were the test pilots. These men were well known by the public, who would flock to air shows to see their favourite 'throw a new aircraft around the sky'. Mostly quiet and unassuming, they faced death every day by flying into the unknown, recording in detail all they felt, heard and observed. Many died – almost one a month at one stage – but their work provided the crucial data for the rapid development of the British jets.

De Havilland

From Light Piston to Fast Jet

Geoffrey de Havilland, frustrated with the engines available to power his light aircraft, engaged his friend and engine designer Frank Holford to find a solution. Holford designed the Gipsy engines, with the cylinders inverted, which reduced drag and improved the pilot's forward vision. The Engine Division established in 1926 went on to produce the famous series of motors that powered the Moth range of aircraft.

De Havilland concentrated on producing aircraft and engines for the RAF and only in 1941 was there an opportunity to explore jet technology. The Halford H1 was a centrifugal engine, as this was cheaper and less risky than an axial flow layout. It differed from Whittle's design in that it adopted a straight-through airflow, which made it simpler to build and operate.

The engine first ran in April 1942 and flew for first time in a Gloster Meteor eleven months later. It was the first jet engine to pass the Air Ministry-type tests in February 1945 and De Havilland became the first British company to build production jet engines. The Goblin, as it was renamed, produced a thrust of 3,100 lbs which made it the most powerful engine available. The Goblin was developed into the Ghost, which was the lightest, the lowest profile and, with 5,000 lbs thrust, the most powerful engine at the time.

In 1961 De Havilland's engine company was absorbed into Bristol Siddeley Engines; five years later, Bristol became part of Rolls-Royce Limited.

Vampire

The Vampire achieved some significant firsts: the first RAF fighter to reach 500 mph; the first jet to land and take off from an aircraft carrier; the first aircraft to reach an altitude of 59,000 feet, and the first jet aircraft to fly across the Atlantic. It was an unsophisticated aircraft without power-assisted controls or radar and only had four 20 mm Hispano guns.

The general design of the Vampire had been agreed on in 1941, but wartime pressures delayed any further work until early 1942. The designers were faced with daunting specifications: managing 100 tons of air per hour through the intakes, the need for a short exhaust, a pressurised cockpit and armament.

The fuselage was a streamlined pod shape made using the tried-and-tested De Havilland composite of a layer of balsawood between two layers of moulded plywood, most famously used to build the

DH Mosquito. Similar to that of a plastic aircraft model, the fuselage was constructed in two halves and then joined together and covered in fabric to provide a smooth finish. In the front was the pressurised cockpit; behind, the Goblin jet engine, metal wings with inbuilt air intakes and, on twin metal booms, the tail surfaces. This unusual design kept the jet exhaust pipe as short as possible, reducing a loss of thrust which was already limited in the early engines. The three prototypes were fitted with tall, pointed tails, but handling difficulties saw these cut down and squared off, eventually to be replaced with the traditional De Havilland tail, reminiscent of the Moth series.

Typical of all the first jets, the Vampire was underpowered, required a long take-off run and was limited by instability in some manoeuvres. The first sixteen production aircraft were set aside as development airframes and all of the aircraft were produced by English Electric as De Havilland was at wartime production capacity. Only about six aircraft flew before the end of the war.

Its light handling characteristics, effective ailerons and elevator made it a very capable little fighter. The characteristic slow throttle response of early jets and possibility of flame out or engine surge, caused by rapid throttle movement, and no restart capability tempered the aircraft's abilities. Performance and handling were found to be optimum at the 370 to 490 mph range; at lower speeds it required firm management, while at higher speeds – as it approached compressibility – it became unpredictable:

> Well, the things I remember about it were that Mach Crit was M 0.78 ... Just over that it started to porpoise madly. The high speed stall had a vicious flick roll which would degenerate into a spin if not dealt with pretty quickly. Moving the stick quickly forward about three inches, stopped the flick roll instantly. [McLeland]

After changes to the cockpit frames, visibility was ideal but the initial layout put switches in awkward places and hid the fuel gauges behind the control column.

Vampire Mk Is entered RAF service in April 1946, but as Mk 3s became available, they were passed on to Royal Auxiliary Air Force squadrons.

Rolls-Royce Nene engines were fitted to three airframes designated Vampire II. They were characterised by the additional intakes behind the cockpit required to feed the double-sided compressor. One aircraft was used in Australia to develop the Nene-powered Vampire F Mk 30.

The second production airframe was used to develop the Vampire III, later designated F Mk 3. It was fitted with a more powerful Goblin 2 engine and integrated wing fuel tanks to increase range and a lowered tailplane.

Development of underwing drop tanks for the Mk III made it possible to explore the range of weapons that could be carried and delivered by a ground-attack version. The resultant FB5 with strengthened square-cut wings, which improved low-level performance, equipped RAF squadrons from 1948. Its ground-attack role was first put to the test during the Malayan Emergency, where 60 Squadron supported troops with rockets and bombs.

The Vampire was relatively simple, as Peter McLeland recalls that there were:

> no bang seats in the single seat Vampires that I flew (or the Venom FB1), so a nice 'Spade Grip' rather than 'Straight Stick'. If you wanted to bale out, you jettisoned the canopy, released your seat

> harness and then rolled inverted ... You kicked the stick forward to bunt the aeroplane away from you (hoping like hell that you would miss the tailplane). [McLeland]

Britain's immediate post-war commitments, which required operations in hot and high conditions, led to the FB9 tropicalised version, upgraded with an air-conditioning unit housed in the right wing root. It was also found that jet engines accelerated slower in the less dense hot air, which gave less lift and required a longer take-off. The more powerful Goblin 3 engine was fitted to the sixteen FB9s exported to Rhodesia (Zimbabwe).

Nightfighter

The RAF issued a specification for a jet fighter to replace the obsolete De Havilland Mosquito nightfighters. De Havilland engineers were convinced they could replicate the success of the Mosquito by using a similar fuselage profile in order to quickly and at low cost build a new cockpit and nose that could be grafted on to a standard FB5.

Although based on an obsolete airframe, the Vampire had proved to be a more stable gun platform with a better range than the Meteor and as the Venom nightfighter was behind schedule it was selected as an interim solution. Its main downfall was the difficulty crew would have escaping in an emergency through the upward folding canopy. Delivered from 1951, the NF10 served for a mere two-and-a-half years.

Some were exported but a number were converted to navigation trainers fitted with new clearview canopies, with quick-release mechanisms.

Trainer

Conversion of the NF10 to a trainer led to another private-venture development: the T11, which was used to train pilots but also as a platform for gunnery and ground-attack exercises. It became a standard stable training platform with the last one withdrawn from RAF use in 1968, but it continued to serve around the world.

In 1954 the Royal Navy ordered a navalised version – the T22 – which was used to train Sea Venom pilots, and which served until the early 1970s.

Sea Vampire

The first pure jet aircraft landings and take-offs took place on HMS *Ocean* on 3 December 1945. Lt-Com. E. M. Brown RNVR performed fifteen take-offs and landings on the vessel over two days. These trials proved the suitability of jets for carrier service: an unobstructed pilot view, the lack of a propeller and a short sturdy undercarriage eliminating swinging or a tendency to bounce on landing. One concern, however, was that the slow acceleration of early jets meant that pilots did not have the instant power to abort a landing.

The Royal Navy selected the Vampire later when more efficient engines and a greater fuel capacity increased their range and led to an order for eighteen Sea Vampire F20 units to replace the De Havilland Sea Hornets. Sea Vampires were fitted with an arrestor hook, larger airbrakes and flaps, and changes to allow for the higher forces of a carrier landing.

Power, fuel and weight ratios were crucial elements when operating early jets and even more so in the naval versions. In an experiment to reduce weight and so increase range and power, three strengthened Sea Vampire F21s and a Hawker Sea Hawk were repeatedly landed wheels-up on a vulcanised rubber and inflated fire hose surface, first on land and then at sea. Practical handling issues precluded further development.

Export

In 1945 the Royal Australian Air Force selected the Commonwealth Aircraft Corporation to build a version of the Vampire. The RAAF's first jet aircraft, the FB30, delivered from September 1949, differed from the British-built examples as they were powered by a CAC-built Nene turbojet. This Rolls-Royce engine was at the time the most powerful engine available and ideal for the hot Australian conditions. The first fifty-six were built to FB5 standards followed by twenty-three FB31 variants with strengthened and clipped wings, which were also fitted to twenty-nine F30s.

For training, thirty-six T33s equivalent to the T11 were ordered in 1951. Further T35s featuring ejector seats and clear canopies were delivered, while the surviving T33s were similarly upgraded to become T35As.

Initial tests in Canada in 1947 showed that the Vampire was unsuited to intensely cold conditions, which affected the undercarriage, oil temperatures and canopies. Politics and the debt owed by Britain decided that the Royal Canadian Air Force had to choose between the Vampire and the Meteor. The former was selected because of its longer range, cockpit visibility and lower price. Delivered from 1948, the aircraft served ten squadrons but was used primarily by the auxiliary units.

In-service problems arose due to: frosting of the canopy, exploding canopies, the management of handling problems at high speed and a basic lack of understanding of jet flight. Once these teething problems had been ironed out, the aircraft proved reliable. Of the eighty-five delivered, forty survived to be withdrawn in June 1956, of which fifteen were sold to the Mexican Air Force in 1960, becoming their first jet fighter.

France received thirty Mk I Vampires from 1948 and then entered into licensed assembly of sixty-seven FB51s, followed by licence manufacture of 120 FB51s using French parts and a licence-built Nene engine. This variant, known as the Mistral, featured modified air intakes and a significant increase in fuel capacity, which required the redesign and strengthening of the airframe. Flying a Mistral, French pilot Jacqueline Auriol set a new world speed record for a female pilot in May 1951. The French De Havillands saw active service on a regular basis in the post-war struggle for Algerian independence.

Newly independent India and Pakistan engaged in regular border clashes, which encouraged the Indian Air force to re-equip with jets. Like the French, they purchased some British-built aircraft but Hindustan Aeronautics Limited licence-built the bulk of the 286 FB52s which first flew in 1952. T55s, mostly Indian built, were used for pilot training. In combat the Vampires were no match for Pakistan's Sabres and a number were lost; however, it was only around 1975 that the surviving Vampires were withdrawn.

From 1949 Switzerland built 100 F Mk 6 Vampires under licence and manufactured the Mk 55 trainer. All these aircraft were fitted with the Martin Baker Mk seats in 1960.

Sweden ordered the Vampire FB50, and the more powerful Goblin 3 FB52 was sold to Finland, Norway, Egypt, Iraq, Lebanon and Venezuela. Twelve FB Mk 5s went to India and twenty-seven to South Africa, while Sud-Est in France assembled another sixty-five. Macchi and Fiat produced eighty FB Mk 52s under licence for the Italian Air Force.

Apart from the United Kingdom, the Vampire served in thirty-two other air forces.

Above: Lt-Com. E. M. Brown RNVR carried out the first jet landing on a ship at sea, the ship being HMS *Ocean*, on 3 December 1945. Sea Vampire F1 LZ551/G, the third prototype, is now part of the 'Carrier' exhibition at the FAA Museum, Yeovilton. (Ken Richards)

Right: A flight of six Vampire IIIs from 54 Squadron were the first jet aircraft to cross the Atlantic, in July 1948, with fuel stops in Iceland, Greenland and Labrador. They were escorted by three Mosquitoes, and supported by two Avro Yorks. Terrible weather put them in danger of running out of fuel and if an engine failed there was no restart option. Bill Wood's aircraft, VT 864, is nearest to the camera. (William Wood)

At the same time a USAF team was attempting to become the first across the Atlantic in F-80 Shooting Stars. The two teams crossed paths in Greenland. The RAF team is seen at Odiham before departure. From left to right: Jeep Hale, Ricky Wright, Frank Wooley, Buck Courtney, Bobby Oxspring, D. S. Wilson-MacDonald (force commander), Micky Martin (Met Mosquito), Bill Wood, Taffy Evans, Roy Skinner. Colin Colquhoun is not present here. (William Wood)

Vampire F1 VF301, built by English Electric in November 1946, served in 595 and 631 squadrons and was then used for training and is seen here on gate guard duty at RAF Debden. This is the only F1 preserved in the UK and is now at Midland Air Museum. (R. A. Scholefield)

Above left: Royal Canadian Air Force Vampire F Mk III c/n EP42313 17021, built by English Electric, delivered in April 1948, and withdrawn from use in October 1956. It was sold in the USA as N6864D and de-registered in 1970. (RCAF)

Above right: A Vampire advertisement that appeared in the March 1956 edition of *Commando.* (Steve McLean)

Vampire FB5 c/n EEP42461, VV217, in 1948, used as a trials aircraft. (R. A. Scholefield)

A formation of SAAF (South African Air Force) Vampire FB-52s, the export version of the FB6 fighter-bomber. (Steve McLean)

Vampire FB52 c/n V0636, SAAF 242. (Steve McLean)

Vampire T11 WZ419 was an early production version. Built in 1952 and fitted with a Goblin 3 engine, it was the first modern fast jet trainer. (R. A. Scholefield)

A plan view of Vampire Mk 55 Trainer U-1208, c/n 868, which was built in 1956 and served with the Swiss Air Force until 1991. (Eduard Bähler)

The other side of HB-RVF, now privately owned and on the display circuit. (Peter Gronemann)

Vampire T55s transiting Bahrain in 1964, on delivery to the Indian Air Force. (Gordon Macadie)

Vampire FB52, c/n V0659, R1835, of the Rhodesian Air Force. Operated by the SAAF as 245, it was loaned to Rhodesia (now Zimbabwe) in 1974. The colour scheme represents the aircraft when it was operational in the Bush War, when all markings were removed. It has been restored to taxi condition by the Old Aeroplane Co., Tyabb, Victoria, Australia. (Barry Broom)

Close-up view of the drop tank and slightly inaccurate dummy 60-lb rockets on R1835. (Barry Broom)

These aircraft flown by the Royal Norwegian Air Historical Squadron are: front, a Vampire T.55, the export version of the T.11; and back, a Vampire FB.52. These aircraft are ex-Swiss Air Force and are in the early 1950s markings of 336 Squadron, Norway's first pure jet fighter squadron. (Royal Norwegian Air Historical Squadron)

Vampire T55-2, c/n 15498, SAAF 277, Rhodesia R4152, ZU-DFH preparing for a flying display at the SAAF Museum, Swartkop, in December 2011. (Steve McLean)

A De Havilland formation with a Mosquito leading, followed by Vampires and, in the rear, a Venom at Warbirds over Wairarapa, New Zealand. The Mosquito FB Mk 26, KA114, originally from Canada, flew in 2012 after a seven-year rebuild. The Vampires are: top, a former Swiss Air Force DH115 T.55, built in 1958, marked in the colours of a No. 75 Squadron Vampire, with a fictitious serial number, NZ5712; below, an ex RAAF Vampire T.35W, ZK-VAM, c/n 4171 A79-649, disposed of in 1970 and restored to flying condition in 1986 and now flown by the Vampire Syndicate. The Venom is an ex-Swiss Air Force FB54 painted in the colours of 14 Squadron RNZAF when they were leased from the RAF and flown out of Tengah, Singapore, in anger against insurgents during the Malayan Emergency, between 1955 and 1958. This example has the pointed nose introduced by the Swiss. (Andrew Caldwell)

Above left: Cockpit of Vampire T11 G-DUSK, XE856, delivered December in 1954 and now at Bournemouth Aviation Museum. (Peter Clarke)

Above right: DH.115 Vampire T55-2 c/n 15486, SAAF 265, sent to Rhodesia as R4221 in 1970. After being retired in 1982, it was imported to Australia in 1988, restored to flying condition in 2001 by Old Aeroplane Co. and is finished in post-Federation RRAF scheme, representing RRAF 119 (R4032). (Phil Vabre)

Venom

Designed as an interim fighter-bomber, the Venom served for longer than expected when the Supermarine Swift failed to meet expectations and the Hawker Hunter was unsatisfactory in this role until later in its career.

Similar to the Vampire in appearance, the aircraft had the significant difference of having a swept wing with tip tanks, the more powerful Ghost engine and bullet fairings on the tailplane. The early versions operated under strict flight restrictions, with broad orange bands painted on the wings to remind the pilots of wing structure problems.

> The flick characteristic was carried over to the first version of the Venom FB1. If anything it was worse. However, they then fixed it completely with a little riveted on, fixed slat about a foot long on the leading edge of each wingtip. This modification absolutely transformed the high-speed stall into a Meteor-like, fierce rumbling judder with no wing drop. [McLeland]

All single-seater Venoms served with overseas RAF squadrons in Germany, Cyprus, the Middle East, Asia and Africa. The first RAF ejection through a canopy was made in 1953 when a wing broke off WE368. This accident revealed that 75 per cent of all Venoms showed a weakness in the rear spar of the wing. Fuel venting from the fuselage into the engine caused a number of fires which went unexplained until Flt Lt Severne managed to land a burning aircraft, jumped out and extinguished the fire, allowing investigators to find the cause.

> Four 20 mm cannons was a lot of firepower in those days and you did carry quite a lot of ammo ... enough for about 30 seconds firing. They made a fabulous loud rattling noise and filled the cockpit with a hazy smell of cordite. Not quite as stable a gun platform as the Meteor but somehow more fun ... Excellent for R/Ps as the rocket rails were close in near the fuselage. You could get very good scores with rockets or cannons in Air to Ground, but the Meteor made better scores Air to Air. [McLeland]

With the problems solved, the Venom performed well in the ground-attack and interceptor roles. It could carry a wide range of stores, was manoeuvrable and provided a stable gun platform. This was proved in combat during the Suez Crisis, in Aden and the Yemen, and in the Malaysian emergency.

The FB4, boasting improved rudder design, hydraulic aileron controls and the ability to carry underwing fuel tanks, was introduced in 1955. These served for a short time only before being replaced by more modern types and they were all withdrawn by 1960.

> It was a great turning aeroplane and if you could cut in behind something then there was no possibility of being out-turned. However, there was a shortage of speed and at Mach 0.86 you went out of control. [McLeland]

Nightfighter

As with the Vampire NF10, De Havilland privately funded the Venom nightfighter and in a similar manner they mated a tandem-seat, long-nose fuselage to the wings and tail of the Venom. Flown in August 1950, it did not impress the RAF as it lacked ejection seats but, as it represented some improvement over the Meteor and Vampire nightfighters, an order was placed for ninety NF2 examples. High accident rates led to only one squadron being operational before the NF2A was delivered with a jettisonable canopy, improved controls and modified flying surfaces.

The introduction of the NF3 in 1955, with its powered ailerons, no outer tailplane and improved radar, saw a total of seven squadrons operate the Venom nightfighter but by early 1958 they had all been replaced.

Only one export order for sixty-two NF51 fighters was received, from the Swedish Flygvapnet. The initial thirty were the first two-seated Venoms to enter service, in December 1952, and were equivalent to the NF2 except that they were fitted with licence-built Swedish Ghost engines shipped to the United Kingdom. The remaining thirty-two units were built to NF2A standard and kits were supplied to upgrade the earlier deliveries. Although out of service by 1960, four aircraft were used as target tugs until 1971.

Sea Venom

Once again the Venom filled a gap between the old and the new when it was selected in 1954 as an interim carrier-based fighter prior to the introduction of the Sea Vixen. After tests with the original Venom prototype, the Royal Navy ordered the Sea Venom FAW20, which incorporated upward folding wings, a V-shaped arrestor hook, catapult pick-up hooks and a clearview canopy that could be jettisoned underwater.

De Havilland continued to improve the design by introducing improved power units, radar, some powered controls and ejector seats, while at the same time strengthened arrestor hooks were fitted after numerous incidences of them breaking. The FAW22 variant became the first Fleet Air Arm aircraft to be armed with heat-seeking air-to-air missiles, the DH Blue Jay, the forerunner of the Firestreak. Armourer Dave Rowland recalls that the ground crew had to be suited and gloved as the ammonia used to keep the electronics cool often leaked. The Royal Navy's Venoms performed well in the Suez Crisis, attacking Egyptian vehicles and infrastructure. Replaced by the Sea Vixen from 1959, the Venom was out of service by July 1961. They were then flown to Abbotsinch, where Dave Rowland remembers removing the Mk III ejector seats to recover the ejection guns for later use.

In 1952 Sud-Est licence-built twenty-five FAW20 versions for the French Aeronavale. Known as the Aquilon (North Wind) 201, these were uprated to include ejector seats and sliding canopies. A further twenty-five aircraft were built in 1954 which incorporated anti-skid brakes, uprated radar and a FIAT-built version of the Ghost 48. The '203' version was fitted with a new American radar; however, this required so much space that the second seat had to be removed. Armed with cannon and the Nord 5103 air-to-air missile, some of the forty built provided ground-attack support in Algeria.

When HMAS *Melbourne* left Portsmouth in March 1956 she was the most modern and best-equipped carrier in her class in the world. On board were thirty-nine FAW53 Venoms – a development of the FAW21. These were the first jet aircraft to serve in a Commonwealth navy. The Venom was selected to maintain commonality with the Royal Navy and was chosen over the Sea Hawk for its all-weather capability, radar and ground-attack role. The last six were finally withdrawn in 1970.

Export

The export version of the Venom proved very successful. With the withdrawal of British forces from Iraq, the Royal Iraqi Air Force was initially equipped with FB1s, which were supplemented with fifteen additional FB50s. Venezuela ordered twenty-two FB54 aircraft in 1956. These served until they were replaced by Hunters in 1965.

The most important export customer was the Swiss Air Force with an order of 126 units with the engines built under licence in Switzerland, as were twenty-four reconnaissance versions of the Venom. It was only in 1983 that the final aircraft were retired and sold off.

The prototype Venom VV612, built as a Vampire FB.5 and converted into a Venom in 1949, seen in flight in March 1950. (R. A. Scholefield)

Sea Venom Mk 20 prototype WK376 engaged in deck trials aboard HMS *Eagle*, May 1952. It was destroyed in a crash in August. (Gareth Bailey)

The Venom armourer's task was pretty tough as all the work was done lying underneath the aircraft. (Peter McLeland)

Venom FB4 at Eastleigh Airport in 1959 being turned around with the fuel bowser behind and an oxygen trolley to the left. (Peter McLeland)

Royal Australian Navy Sea Venom FAW53s over Cubi Point, Subic Bay, the Philippines, in 1963. (Kim Dunstan)

HMAS *Melbourne* enters Milford Sound, New Zealand, in February 1957, seen here with seven Sea Venom FAW53 fighter-bombers and one Fairey Gannet on her flight deck, while crew appear to be laying out the arrestor wires. (George Self via Kim Dunstan)

Above left: A 1961 view of the Bristol Sycamore plane guard landing behind Sea Venom FAW53 801, which has just landed aboard HMAS *Melbourne*. (Kim Dunstan)

Above right: This Sea Venom FAW 53 has failed to hook the arrestor wire on HMAS *Melbourne* and is heading over the side in what is known in Fleet Air Arm parlance as a 'bolter'. The pilot would open the throttle, gain height and reposition for another attempt at landing. (Kim Dunstan)

Sea Venom FAW22 c/n 121132, XG731, aboard HMS *Centaur* in the Gulf of Aden in 1959. It was sent to the Tain ranges after 1968. (Owen Edwards via Rob Pennycook)

Sea Venom FAW21 c/n12715, WW141, FAA LM/651, delivered in 1955 and seen here awaiting scrapping with many others and Sea Hawks at Abbotsinch in 1961. (Gordon Macadie)

Classic Air Force's Venom FB50, c/n 824 J-161, G-VENM, was built in 1955 in Switzerland by Pilatus under licence from De Havilland as an FB50 with the construction number 824. Used for high level photo-reconnaissance duties and later for training, it now flies in the markings of Venom FB1 WK436 of 11 Squadron as it appeared in 1955. (Tony Hisgett)

Swiss Air Force Venom FB50 Mk 1R c/n 840 J-1612, HB-RVA. In the 1970s it was decided that all Swiss military aircraft were to be fitted with an IFF transponder. This required a new nose design which was rather more pointed than the original. This nose was also fitted to the surviving Vampires. (Propfreak Collection)

Classic Air Force's Venom FB 50, c/n 752 J-1542, G-DHVM, followed the same career as stablemate G-VENM. Its extended nose was replaced with the standard unit in 2000. Representing 'WR740' of 208 Squadron, RAF, in the 1950s, it is seen here showing the smoke plume from the starter cartridge. (Classic Air Force)

Comet

The scene is London Airport; the time is 14:45 hr local time on May 2nd, 1952. Airliners of many nations crowd the apron, their cabin-tops white against an inky sky. Among them, out of sight of the many observers at windows and on terraces, is BOAC's De Havilland Comet G-ALYP. Today it carries for the first time fare-paying passengers. With the full complement of 36 installed, and his initial cockpit checks completed, Capt AM Majendie, Flight Captain of the Comet Fleet, is ready to start the four Ghost turbojets. So began the first scheduled jet flight to Johannesburg, where the Comet arrived 23 hr 38 min later – two minutes ahead of schedule. [*Flight Global*]

It was three years earlier, on 27 July 1949, that De Havilland chief test pilot John Cunningham had flown the DH106 on its maiden flight. After some test hops for the press, Cunningham took the plane to 10,000 feet and then flew it along the runway at 100 feet for the benefit of the team who had designed and built the world's first purpose-built jet airliner. Its four rather underpowered Ghost turbojet engines, fitted in pairs within the wings, were capable of taking it to 40,000 feet at nearly 500 mph, carrying thirty-six First Class passengers.

The Brabazon Committee, chaired by the First Baron Brabazon of Tara, John Moore-Brabazon, was formed in December 1942 to investigate the future requirements of British civilian airlines. It was agreed that a jet-powered aircraft was the best opportunity for Britain to take the lead in airliner development. In 1946 De Havilland designer Ronald Bishop began work on what was to become the Comet. An experimental design – the tailless DH108 – contributed greatly to the layout of the Comet but at the cost of the life of Geoffrey de Havilland Jr.

The Comet 1 introduced a number of technical features, such as: streaming radio aerials replaced by suppressed aerials housed in fuselage fittings, cloud- and collision-warning radar and power-operated flying controls. The metal skin surfaces were bonded to the frame rather than riveted, providing a lighter, smoother finish and thin wings. It was fully pressurised to provide for passenger comfort at its operating height of 40,000 feet.

BOAC had placed an order for sixteen aircraft in 1946, which encouraged De Havilland to go into production before a prototype had even flown. De Havilland provided a luxurious passenger experience at heights and speeds only previously experienced by military pilots and required no changes to the airport infrastructure of the time. All of this was achieved within two and a half years, on time and within budget – an outstanding achievement.

This beautiful, innovative aircraft contributed to the future of air travel in the most valuable but tragic manner. There were three unexplained crashes in quick succession: the first in May 1953, when a Comet 1 G-ALYV crashed at Calcutta six minutes after take-off, followed on 10 January 1954, when G-ALYP broke up off the island of Elba 20 minutes after take-off. On 8 April 1954, the last of the three, G-ALYY, crashed into the sea off Naples.

The fleet was grounded and another airframe was put into intensive ground-based pressurisation tests. The results showed evidence of metal fatigue around the distinctive square windows, which would have caused sudden depressurisation and resultant structural failure. Comets were withdrawn from service and modified or scrapped and the development of the Comet 2 was put on hold.

Once the metals had been developed that could manage the high temperatures and pressures of the axial flow engine, Rolls-Royce created the Avon engine, producing 46 per cent more thrust, which in August 1953 powered the first Comet 2. With the new engines, passenger capacity rose to forty-four, and the increased range opened up new routes, which encouraged eight other airlines to place orders.

Following the Comet 1 disasters, fifteen of the twenty-five Comet 2s in production were modified with elliptical windows and heavier-gauged metal skins. Airline orders were cancelled and all but four were delivered to the RAF's 216 Squadron, which became the first transport squadron to operate jets. The squadron performed sterling service in the RAF as general transports and served as flying laboratories during the development of missiles in the United Kingdom and the United States, being withdrawn only in January 1975. RAF Comet 2 pilot Ollie Alderson wrote that he

> disliked its handling qualities as it lacked a yaw damper and snaked around on finals like a demented rabbit. The flight deck was archaic and I was amazed to find that the stand-by compass was the same one that I had last used in the Tiger Moth, the type that had a magnetic needle floating around in a bowl of spirit. The flight deck was festooned with levers and wheels that seemed to me positively pre-historic. [Alderson]

The Comet 3 announced at Farnborough in September 1952 was a larger and much-improved version of its predecessors; with more powerful engines, it could carry seventy-six passengers. Importantly, Pan American Airlines placed an order in October 1952 for three and secured options for seven units of the new airliner.

The only Comet 3 built, G-ANLO, was used to develop the Comet 4. It completed the first around-the-world jet flight in December 1955; then, with the RAF, as XP915 it was used for blind-landing experiments, in trials to replace maritime patrol Shackletons. After it was withdrawn from use it became the Nimrod mock-up airframe, finally being scrapped at St Magwan in 2001.

Challenges of the Jet Age were faced, overcome and incorporated into the Comet 4, which served commercially for twenty-two years and with the RAF until 1997. The Rolls-Royce Avon units provided a longer range, higher speed and greater carrying capacity that doubled the number of passengers carried by the Comet 1.

Comet's great rival, the Boeing 707, entered service within nine days of the Comet 4. However, it was larger, a more modern design and the greater wing sweep provided better efficiency and range. The 707 was not without its problems, which led BOAC to order twenty Comets instead, and in October 1958 it became the first airline to offer scheduled Atlantic crossings by jet airliner.

To meet the requirements of short-range operators the Comet 4B introduced a stretched fuselage and shorter wings, going into service with British European Airways in 1958. A further development was the 4C, which was, with its even longer fuselage and longer wings, aimed at the long-haul carriers. This variant had an excellent power-to-weight ratio that endowed it with quite a spectacular climb immediately after take-off, a feature that was great fun to demonstrate at low take-off weights. We carried about 100 passengers, not as luxurious seating as the Comet 2 but perfectly adequate for normal passenger tasks. [Alderson]

However the Comet had fallen behind in design and engine technology, while De Havilland's management had failed to grasp fully that the world was preparing for large jets.

Dan-Air's Comet 4C G-BDIW closed the Comet era with the last commercial flight in November 1980, ending nearly thirty years of passenger service by the type.

The scrapheap seems an unfitting end for such an aeroplane.

Chris Kjelgaard

The prototype Comet, G-ALVG, in December 1949. It was broken up in July 1953. (R. A. Scholefield)

The first production Comet 1 before delivery to BOAC. This aircraft operated the world's first scheduled jet flight in May 1952. Beyond are the prototypes, G-ALVG and G-ALZK (far). (R. A. Scholefield)

Demonstrating at the SBAC in 1954 is Comet 2 G-AMXD, which later served with the RAE as Comet 2E XN453. (R. A. Scholefield)

Comet 4B c/n 06435, G-APMD, built 1959 for BEA and seen here in Dublin in the late 1960s. It was withdrawn from the Dan-Air fleet in 1975. (Malcolm Nanson)

Comet 4B c/n 6437, SX-DAK, was built in 1960 and delivered to BEA, leased to Olympic and ended up with the MOD until it was withdrawn in 1978. It is seen here taxiing at Heathrow in the mid-60s. (Malcolm Nanson)

Comet C4 c/n 6471, seen here at Shannon, was delivered in 1962 to 216 Squadron RAF as XR399. In 1975 it was sold to Dan-Air. It was the last Comet to fly in commercial colours when it flew to the National Museum of Flight in Scotland in 1981. (Malcolm Nanson)

In the end, the Boeing 707 conquered the market and Comets were slowly withdrawn from service. This photo shows early morning BOAC arrivals at the Oceanic Building with 707-436 G-ARRB, Comet 4s G-APDD and G-APDK in the background. (R. A. Scholefield)

In full RAE colours, DH106 Comet 4 XS235 (G-CDPA) *Canopus* blasts down Bruntingthorpe's runway on the 2012 Cold War Jets Day. (Alan Wilson)

A direct descendant of the Comet flew until 2011 in the form of the Nimrod, and there is a clear family resemblance between A&AEE Comet XS235 and Nimrod XV239. (Rob Finch)

Gloster

Britain's lead in the post-war Jet Age was provided by the Meteor, which remained until the Americans produced the North American F86 Sabre in October 1947 and the Russians the MiG-15 in the December. These were the first aircraft that took advantage of flight research data captured from world-leading German aeronautical engineers. On the other hand, Gloster had based their design on a conventional piston-engine layout and it could not be developed aerodynamically any further.

The Meteor was the backbone of the RAF until at least 1956 and in the early days it was the flagship of the much-admired British aviation industry. Meteors flew to all the corners of the world breaking records and were most often the first jet-powered aircraft to visit a country. Test pilot William Waterman said that 'despite its faults, the Meteor was far in advance of anything the rest of the world could offer'. [Waterton, 38]

Rapid change

The Jet Age was characterised by experimentation and quick developments, which resulted in the regular production of new versions of an aircraft. Before the end of the war the Meteor F Mk IV flew with Derwent V engines, a shorter wingspan, high take-off and landing speeds and was the first of the Meteors to be pressurised.

> The Meteor was a new experience waiting to hit a pilot who had never flown anything more powerful than the Harvard. Its twin Derwent engines packed a terrific punch compared to the Pratt & Witney Wasp; each engine developed around 3,500 pounds of static thrust and fuel for thirty-five minutes of flight time, if you were lucky. [Alderson]

Pilots liked the Meteor, but the early models

> suffered from the appalling design problems caused by the relight buttons being located on a panel on the right-hand side of the office; meanwhile, the HP fuel cocks were levers on either side of the seat. The contortions required to restart an engine were legendary, rudder application to keep the beast straight, one hand gripping the stick and the other groping to find the relight button on the right hand side of the cockpit. Now we needed a third hand to feed on the HP fuel cock. Impossible you might say, and you would be absolutely right, the only way was to wrap one leg around the stick, using the resultant free hand for the relight and the other for the HP cock, while the leg holding the asymmetric load ached and shuddered in protest. [Alderson]

By 1947 it equipped thirty-one RAF and Auxiliary squadrons, but was already falling behind its rivals. An improved design resulted in the F8, the most powerful and efficient of all Meteors, which became the RAF's primary fighter from 1950 to 1955. Ballast was removed from the nose, which was lengthened to hold additional fuel, and to compensate a new tail unit was required, which came from the experimental Gloster E1/44 fighter, and all F8s had the new, more angular and taller tail assembly. They were also powered by Derwent 8 engines and were fitted with ejector seats and a blown canopy, providing better visibility.

> The Meteor's handling qualities were truly excellent; it was the most forgiving of aeroplanes and was a joy to fly. It performed quite happily at Mach 0.74 with only marked control stiffening causing any great problems above that speed. [Alderson]

There were numerous losses through what was known as the Phantom Dive, when on finals the aircraft would simply drop out of the sky, particularly evident in the T7. Hal Taylor, the driving force in the preservation of Meteor *Q-Queenie*, found that the port wheel locked down before the starboard one which, combined with extended airbrakes, created asymmetrical flying conditions and a rapid rate of descent. He concluded that the large glass cockpit of the T7 masked the flight surfaces at low speed and insisted that pilots ensure the airbrakes were not deployed before they lowered the undercarriage.

Records

The new technology opened up a world of firsts and new techniques. The High Speed Flight was formed in June 1946 to set a speed record over a course laid off the south coast between Rustinton and Angmering. Nimonic, a nickel-based, high-temperature metal compound developed in Wiggen and used for the turbine blades, allowed the engine to operate at higher temperatures and produce more power. The two aircraft to be used were highly prepared and polished and had 3 feet removed from each wing to provide better stability at speed. Group Captain Teddy Donaldson in EE549 achieved a world-best time of 616 mph (991 km/h) on 7 September 1947.

Another international speed record was set in April 1950 by Squadron Leader Janusz Żurakowski when he flew London–Copenhagen–London in a standard F8 (VZ468). The Danes were suitably impressed and purchased the type soon after.

Żurakowski also developed the first new aerobatic manoeuvre in twenty years, the 'Zurabatic Cartwheel'. At the 1951 Farnborough Air Show Żurakowski used the asymmetric behaviour of the Meteor to achieve a seemingly stationary vertical cartwheel. Climbing to 4,000 feet, slowing and then cutting one engine caused the aircraft to pivot; as it reached the vertical, the other engine was throttled back, resulting in a full 360 degree rotation.

An endurance record of 12 hours 3 minutes was achieved by pilot Partrick Hornidge of Flight Refuelling Ltd. Refuelled ten times by a Lancaster tanker, he covered 3,600 miles (5,800 km) on 7 August 1949. This proved that jets, notoriously fuel inefficient and short on range, could remain aloft for long periods.

Nightfighter

Gloster proposed a Mosquito nightfighter replacement based on the two-seater Meteor T7. The NF11, built by Armstrong Whitworth, used the tail of the T8 and the wings of the F3 and an extended nose for the radar, which forced the relocation of the cannon to the wings outboard of the engines.

The Meteor nightfighter was supplied to the RAF from 1951 and tropicalised as the NF13, serving during the Suez Crisis. Ollie Alderson, who had three operational tours on nightfighters, remembers that

> the Armstrong Whitworth Meteor NF11 had the same awful glasshouse canopy as that fitted to the Meteor T7. The nose was elongated to accommodate the Airborne Interception Radar, in this case, the AI Mk10. We still did not have the benefit of ejector seats and leaving the aircraft in an emergency was something of a gamble as the tail tended to get in the way. The navigator occupied the rear seat and spent his life with his head stuck in a visor gazing at two radar scopes, one providing a horizontal picture with range markers (the B scope), the other provided a vertical picture (the C scope). The maximum range available was usually around six miles and rarely more than ten. The navigator's hands were kept busy twiddling knobs controlling the gain level while he shouted horizontal and vertical commands that were intended to allow the pilot to adjust the flight path in such a way that he would eventually be able to obtain a visual sighting on the target. [Alderson]

During his time in West Germany, Alderson noted that they flew air-to-air film, air-to-ground firing, and low-level cross country sorties and in between became very proficient at formation flying.

With the advent of more powerful US-built radars, an NF12 was produced with a 17 inch (43.2 cm) longer nose, an enlarged fin, new Rolls-Royce Derwent 9 engines and strengthened wings. The final Meteor fighter was the NF14 which, although deliveries only began in 1953, was already being replaced in 1956 by the Gloster Javelin. Fourteen of these were converted to the NF(T)14 navigation trainers, finally being retired in 1965.

> Looking back now I can see that the instructors were as much in the dark as the students with regard to how best to fly a jet aircraft.
>
> *A Royal Air Force Pilot Remembers,* Sqn Ldr O. W. Alderson

On Foreign Service

In 1946 single F Mk 3 examples were loaned to the air forces of Australia, Canada, New Zealand and South Africa, but no sales were made. However, the Meteor in its various forms came to dominate the new world of jet fighter aircraft.

Europe

Belgium was the largest foreign user, operating 355 Meteors from 1948 when the first TMk7s were delivered, followed by the F Mk 4s and then F Mk 8s. Belgium also followed the United Kingdom in replacing their Mosquito nightfighters with Meteor NF11s, which were only replaced in 1959.

France had evaluated the F Mk IVs in 1948 and in 1951 two T Mk 7s, and then selected the NF11 to replace its aging Mosquito nightfighters. These were flown in combat in Algeria, while four NF11s and later two each of the NF13 and NF14 were used extensively as test beds and contributed to the Concorde programme, finally being retired in the 1980s.

The Netherlands operated a large fleet of sixty-four F Mk IIIs, forty-five T Mk 7s and 160 F Mk 8s. From July 1948 Fokker produced Meteors under licence at Schiphol. Meteor F Mk 8s were replaced by Hawker Hunters from 1956.

Twenty F Mk 4s were ordered by the Danish Navy, entering service in 1949 and a year later they formed part of the new Royal Danish Air Force, when the air units of the army and navy merged. The RDAF replaced the F4s with twenty F8s, supplemented by T7s and twenty NF11s.

The Middle East

Political problems delayed deliveries of Egypt's twelve F Mk 4s and three T Mk 7s for almost a year until 1949. Only four of a further order for twelve were delivered before the remaining eight were embargoed and sold to Israel and Brazil. The eight were eventually delivered in 1955 with three more T Mk 7s and six NF Mk 13 nightfighters. During the Suez Crisis the RAF, Egypt and Israel all flew Meteors and, although no Meteor-to-Meteor combat was recorded, a number of Egyptian examples were destroyed whereas one is reported to have damaged a Valiant bomber.

Syria's order for two T Mk 7s and twelve F Mk 8s in January 1950 was also affected by the arms embargoes. Only the F Mk 8s had been delivered by December 1952 when another arms embargo prevented the delivery of the T Mk 7s. Seven surplus RAF F Mk 8s, two FR Mk 9s and six NL Mk 13 aircraft were delivered later.

Finding the United States unwilling to sell jet aircraft to Israel, the Israelis looked to Britain, who had sold jets to their neighbours. The British believed that supplying Israel would maintain the balance of power and delivered four T Mk 7s and eleven F Mk 8s, and in 1955 seven FR Mk 9s and two more T Mk 7s, and the following year six NF Mk 13s. The FR Mk 9s were stripped of their reconnaissance equipment and used as fighters. Additional hybrid T Mk 7s were sourced from Avions Fairey, who had converted F Mk 4s and added F Mk 8 tails. Realising that the MiG-15 out-classed the Meteor, the Israelis did not use them as fighters but rather in a ground-attack role.

In Operation Tarnegol ('Rooster') an NF Mk 13 intercepted and destroyed an Ilyshin Il 14 carrying high-level Egyptian officers en route from Syria to Egypt on the eve of the Suez Crisis.

South America

Taking advantage of a post-war positive balance of trade, Argentina ordered a range of British aircraft in May 1947, including 100 Meteors. Argentinian pilots began their training at Morten

Valance – first on Avro Ansons and Airspeed Oxfords – to become familiar with multi-engine aircraft before they moved on to the Meteor. There were no two-seater aircraft and, as Bill Waterton noted,

> We removed the canopies from Mk IIIs and sat cowboy fashion on the ammunition boxes behind our pupils yelling instructions above the roar of the engines, and signalling by hand. [Waterton, 110]

The first aircraft were delivered in July 1947 and Bill Waterton carried out the first jet flight in Argentina aboard I-005 on 11 July.

In 1958, the Glosters, as they were known, were re-assigned as fighter-bombers, fitted with bomb racks and rocket rails, and their codes were changed from I (Interceptor) to C (Caza – 'Hunting'). Service continued until December 1970, when twelve aircraft performed a final farewell formation. Two examples continued flying until March 1971.

After the war Brazil was the second largest air force on the sub-continent but it was all piston-engine based and soon fell behind the rest of the world. It could not afford the more expensive American jets and so traded 15,000 tons of cotton for 60 F Mk 8s (F8) and 10 F Mk 7s (TF7). The F8s, the first of which flew in February 1953, were configured for interception and ground-attack roles. The fitted canopies on all F8s had a metal strip at the back which caused the canopy to explode when, after being in the Brazilian sun, an aircraft reached a high altitude and colder temperatures. This was resolved by fitting all clear canopies. The Meteors were almost exclusively used in close air support, which led to their retirement in 1968 due to main wing spar fatigue arising from the stresses of low-level, high-speed flight. The last TF7 was withdrawn in 1971, while a single F8 was put together from parts and used as a target tug until April 1974.

Only equipping one squadron in 1954, the twelve FRMk9s in Ecuador were in service until the late 1970s, when they were replaced by the SEPECAT Jaguar.

Australia

When the RAAF was sent to Korea in support of Allied forces in 1950, ninety-four F Mk 8s and nine T Mk 7s were ordered. Serving with 77 Squadron from June 1951 to the end of hostilities in July 1953, it took a while and some serious losses for the Meteor to find its niche.

Initially used in as an interceptor, it soon became obvious that the MiG-15 was superior. Air-to-air encounters with the MiG had left three pilots dead and one a prisoner and with the arrival of additional USAF swept-wing Sabres, the Meteor were moved to a ground-attack role in January 1952.

Low-level, high-speed flying was hazardous and hard on the aircraft, but the Meteor was at home at lower altitudes and could outclass the enemy, shooting down two MiGs. Losses were still high as a result of ground fire; in total thirty-eight personnel were lost, seven pilots captured and fifty-four Meteors lost to all causes. RAAF 77 Squadron flew 15,000 Meteor sorties, achieving outstanding results destroying enemy equipment and facilities, but most importantly, completely disrupting supply lines. A significant factor in this success was an innovative rocket containing napalm first used in February 1952.

The final exported Meteors were two NF14 aircraft 'smuggled' out of the United Kingdom in 1969: G-ASLW (WS829) and WS804, registered as G-AXNE, were sold by Rolls-Royce and Aeroflight, respectively, and were flown to Faro in Portugal, where they were fitted with four 20 mm cannon and bomb racks for their use with Biafran forces against Nigeria. G-AXNE had successfully flown from Faro via Madeira and Dakar to Bissau when G-ASLW attempted to fly directly to Bissau omitting Dakar. Flying far out to sea and misjudging the Meteor's range, the pilot found he had run short of fuel and ejected close to a ship, which picked him up. By this time the British government had realised the proposed final destination of the aircraft and demanded that the Portuguese authorities impound G-AXNE. The aircraft was towed to a remote part of the airfield where she remained, deteriorated and disintegrated.

> The Meteor, which was built in thirty-four sub-types with twenty-one different power units, helped change the whole course of military aviation in many countries.
>
> D. N. James [James, 127]

Above: A Gloster Meteor F.3 of Royal Air Force 616 Squadron in flight, 18 December 1944. (Phil DeFer)

Right: Built in 1946, Meteor F.4 7008M, EE549, set a new world speed record of 616 mph on 7 September 1946, flown by Group Captain E. M. Donaldson of the RAF High Speed Flight at Tangmere. It is seen here on Horse Guards Parade in September 1968 and is now at Tangmere Military Aviation Museum. (Malcom Nason)

Two Mk IIIs, EE337 and EE 387, were fitted with arresting hooks and reinforced undercarriages for trials aboard Royal Navy carriers HMS *Illustrious* and *Implacable* in 1948. Here HMS *Illustrious* deck crew are releasing the arrestor wire from EE387. (Frank Broughton via David Page)

Egypt was the first Middle Eastern country to acquire jet aircraft but an arms embargo delayed delivery of the first two of twelve F Mk 4s and one of six T Mk 7s until October 1949. Seen here prior to delivery are Mk 7 serial 1400 and Mk 4 serial 1401. (Jet Age Museum)

Ex-RAF Lancaster B Mk 3 PB972, G-33-2, was used for early development probe-and-drogue refuelling tests. A Meteor F Mk3, EE397, is receiving over Poole in August 1949. (R. A. Scholefield)

Meteor T Mk 7 VW481, delivered in 1949, was on the final leg of single-engine approach to Little Rissington on 5 April 1950 when the nose rose and turned 90 degrees and it crashed from 500 feet up and 1,100 yards short of the runway; the pilot was seriously injured. The crash was possibly the result of the 'Phantom Dive'. (Russell Adams via Jet Age Museum)

Gloster's private venture ground-attack Meteor (GAF) *Reaper*, G-AMCJ, G-7-1, a development F Mk 8, was designed to carry up to twenty-four 96-lb rocket projectiles. (Russell Adams via Jet Age Museum)

G-7-1, armed with four 1,000-lb bombs and fitted with two 100-gallon wing-tip fuel tanks, approaches the Severn railway bridge in August 1951. (Russell Adams via Jet Age Museum)

During 1948 an F8, WA820, was fitted with two Armstrong Siddeley Sapphire turbojets and on 31 August 1951 it achieved a number of consecutive time-to-altitude records, finally reaching 39,370 feet in 3 minutes and 7 seconds. (Russell Adams via Jet Age Museum)

Manchester's Ringway Airport in July 1952, where a 613 Squadron Meteor T.7, WF778, awaits another training flight. Starter units for Vampire FB.5s can be seen and a Vampire is being refuelled in the background. (R. A. Scholefield)

Known as the TF7 in Brazil, this example, 4304, c/n G5-444881, was photographed in 1953 while ten Brazilians were under instruction at Morten Valance. This aircraft was scrapped in 1967. (Russell Adams via Jet Age Museum)

Airframe workshops at RAF Halton, 1955. (Brian Goodwin)

In July 1955 the Central Flying School aerobatics team took part in the French National Air Meet at Le Mans. T7 WL373, second right, was flown by Flt Lt D. Hall. (Russell Adams via Jet Age Museum)

The inverted leader formation, performed by the Central Flying School's 1956 aerobatic team. (Russell Adams via Jet Age Museum)

A similar formation seen from above. (Jet Age Museum)

A Royal Australian Air Force Gloster Meteor F.8 fighter, A-77-570 from No. 77 Squadron, at K-14 Kimpo air base during the Korean War. (USAF)

Built by Gloster, this aircraft served with the RAF as VZ467, ending as the famous target tug *Winston*. It then became privately owned and was on the display circuit until purchased by Temora Aviation Museum in Australia, where it is now marked as MiG-killer A77-851 (VH-MBX). (Michael Wignall – naemickpics.com)

Meteor T.7 WA591 (G-BWMF) was built by Gloster Aircraft at Hucklecote, Gloucestershire, in mid-1949 and delivered to 226 OCU at RAF Stradishall in October. It suffered four Cat. 4R accidents and had to be returned to Gloster for repairs. It suffered a lesser accident before it was declared obsolete in 1965 and became a gate guardian at Cheshire. It was rescued in 1995 by Meteor Flight and in 2007 Air Atlantique's Classic Flight agreed to help fund the restoration and subsequent operation. (Classic Air Force)

A pair of experimental F.8s. *Flight* of 27 May 1955 records that VZ460 of the Central Fighter Establishment was used for tests with rocket projectiles and bombs, while VZ473 was fitted with an FR9 nose at some stage, ending its days in 1962 at the MOA Fire School at Stansted. (Warbird Exchange)

This is the 1953 prototype Meteor FR9, VW360, which had a lengthened forward fuselage to accommodate an F24 camera, which could take pictures through one of three windows in the nose. (Russell Adams via Jet Age Museum)

While the FR9 retained its cannon, the PR10 was unarmed, designed for high-altitude operations with an extended wing and the elliptical tail unit of the F4. VZ620, seen here in 1951, was one of fifty-nine manufactured. (Russell Adams via Jet Age Museum)

Meteor NF11 WD587, delivered in April 1951, was known to have served with the Radar Research Establishment at Defford Airfield in Worcestershire, the main station in Britain for the development of airborne radar during and after the Second World War, before it was scrapped in February 1958. (Russell Adams via Jet Age Museum)

Gloster Meteor NF.11 WM167 (G-LOSM) was built under licence from Gloster by the Armstrong Whitworth Company at Baginton (Coventry) airport in 1952. It was modified in 1961 to TT.20 target towing configuration and used as such until 1975. Bought privately and converted back to NF11 standard, it was owned by a number of organisations until it was purchased by Classic Air Force, Coventry. (Classic Air Force)

Over the years G-LOSM has featured in a variety of documentaries and has taken part in a number of important flights – including carrying the ashes of jet-engine designer Sir Frank Whittle. (Classic Air Force)

Belgian Meteor F.8 EG-10/MN-S of 350 Squadron of the 1st Wing, based at Beauvechain in the early 1950s. (Andre Bar via Belgian-Wings.be)

Fokker Aircraft-built Meteor F.8 EG-121 of the Belgian Air Force at Blackbushe Airport, 1955. The F Mk 8s equipped twelve squadrons until they were replaced by Hawker Hunters. (Ronnie Bell)

This picture of Armstrong-Whitworth Meteor NF14 WS830, of 46 Squadron, accompanied by three NF12s highlights the 17-inch nose extension to accommodate further equipment and the new bubble canopy. This aircraft served from 1954 to late 1957. (Russell Adams via Jet Age Museum)

Argentina purchased fifty low-hour ex-RAF machines and fifty new G41G MkIVs, an aircraft which had a shorter wingspan than the Mk IV and was powered by the Rolls-Royce Derwent V. (Juan Carlos Cicalesi)

In 1958, the Glosters, as they were known, were changed from interceptors to fighter-bombers, fitted with bomb racks and rocket rails, and their codes were changed from I (Interceptor) to C (*Caza* – 'Hunting'). Over time, they were camouflaged. (Juan Carlos Cicalesi)

Argentinean Meteors were used in anger for the first time in a revolt against Perón in the summer of 1955, when one shot down a rebel AT-6 Texan. In a second successful autumn revolt, rebels seized four Glosters, which they marked with a cross on a V with the legend '*Cristo vence*' ('Christ wins'). A further uprising in April 1963 saw the Glosters attacking the naval base at Parque Chacabuco. (Juan Carlos Cicalesi)

Denmark purchased nine T7 trainers, of which only two survived to be retired in 1962. (Russell Adams via Jet Age Museum)

World-renowned and pioneering photographer Russell Adams, seen here about to board a Javelin at Morten Valance in February 1962. (Jet Age Museum)

Hawker

New methods of operation were required by the Royal Navy to manage heavier aircraft with high landing speeds. The mirror landing aid, the steam catapult and the angled flight deck were all introduced and then adopted by the navies of the world.

Sea Hawk

The Sea Vampire and the Attacker suffered from a very short range, while the Attacker's tail-dragging layout was difficult to manage on deck. Naval aircraft are required to remain airborne for longer than their land-based counterparts and to have excellent low-speed handling characteristics.

Available centrifugal engines were stubby and round, requiring a short exhaust pipe to reduce the loss of thrust. Hawker's designer, Sydney Camm, devised a novel solution to the problem. He placed the Rolls-Royce Nene engine in the centre of his design. It was fed by two air intakes placed on either side of the wing root, but the real innovation was the bifurcated jet pipe, which split the engine's exhaust gases so that they exited at the trailing edge of each wing.

The pilot was placed in the nose, providing excellent visibility, and the engine positioned near the centre of gravity; the result was great manoeuvrability. Moreover, the flow of air into the engine via the leading and trailing edges of the wing provided space for additional fuel tanks in the front and rear of the fuselage, which solved the range problem.

Initial drawings to meet RAF requirements were developed in 1944, but by October 1945, the RAF was only interested in their new Meteor. This, along with the post-war cancellations of the company's piston-powered aircraft, led the designers to modify the P1040 for carrier-based operations and submit the idea to the Royal Navy.

Powered by the Rolls-Royce Nene or RB41, built in only five months and doubling the thrust of the Whittle engine, the new Hawker first flew in September 1947. A year later the first navalised version flew, with folding wings, catapult launching and arrestor hook.

After successful trails 151 Sea Hawks were ordered in September 1949. The outbreak of the Korean War gave the new Hawker 'Super Priority' status and production was moved to fellow Hawker Siddeley Group company Armstrong Whitworth, allowing Hawker to concentrate on the equally important Hunter.

The F1 entered service with 806 Squadron in 1953 with powered aileron controls, a strengthened structure, an increased range of weapons that could be delivered and, with the FB5 and FGA6, an uprated Nene power unit.

At War

The Sea Hawk's primary role was to cover the fleet, with a secondary role of reconnaissance and ground attack. The FB3 was strengthened to carry external weapons and it was this version that saw action in the 1956 Suez Crisis when Britain and France attempted to maintain control over the all-important Suez Canal.

Delayed shipments of Vampires and Meteors, substandard training and the Israeli conflict made it difficult for the European powers to arm Egypt and therefore the new Egyptian government looked to the East for military equipment.

New, more powerful aircraft coupled with the increased support for Israel by France, followed by the withdrawal of funds for the Aswan Dam, led the Egyptian government to nationalise the Canal. Operation Musketeer, the plan to invade the Canal area and to take back control, began with the arrival of a joint Royal Navy and French fleet in the eastern Mediterranean at the end of October 1956 and a large number of fighters and bombers were flown to bases on Cyprus. While the combined British, French and Israeli air forces outweighed the Egyptians, none of their aircraft could match the Russian MiG-15.

Militarily the campaign was rated as a success, one in which the Sea Hawks were used to attack ground targets using their cannons, which they did with great efficiency, but politically it was a disaster as the United States forced the three antagonists to withdraw.

The elegant Sea Hawk proved to be a dependable military aircraft but rapid technological developments meant that it was to become obsolete after a very short service life. The Fleet Air Arm (FAA) began to phase them out in 1958, and the last first-line squadron was disbanded in December 1960, with the type finally being withdrawn in 1969.

In Foreign Service

Germany's land-based Bundesmarine ordered sixty-four Sea Hawks, split equally between F Mk 100 dayfighters for the maritime strike role and F Mk 101 all-weather fighters. These were modified to carry the Ekco Type 34 search radar on one of the underwing pylons, which required a 38 cm (15 in) taller tailfin to compensate for the large pod.

Only fifty-three had been delivered by the time they were replaced by the Lockheed F104 Starfighter in 1963. Their short German service life over, twenty-eight were sold to the Indian Navy.

The Dutch Koninklijke Marine bought thirty-two Sea Hawk F Mk 50 aircraft for use on their carrier, the *Karel Doorman* (ex-HMS *Venerable*). When the NATO mission profile changed to anti-submarine warfare, the Sea Hawks were taken out of service in 1964 and the *Karel Doorman* was sold to Argentina.

By 1959 the production line had been closed and was being dismantled when the Indian Navy selected the Sea Hawk to equip their carrier INS *Vikrant* (ex-HMS *Hercules*). They ordered fourteen new-build Sea Hawks, which led to the jigs being hastily reassembled and twenty-two ex-Fleet Air Arm aircraft being refurbished to Mk 6 standards. This order was supplemented in 1965 with ten Mk 100s and eighteen Mk 101s from Germany that were refurbished by Armstrong Whitworth. In the 1965 Indo-Pakistani war, the Hawks were used for ground attack while in the 1971 war they operated from the carrier and were successful in sinking Pakistani gunboats and freighters.

Over 10,000 catapult launches took place aboard INS *Vikrant* and only one went wrong. On 4 March 1976 the catapult launch bridle on Commander Peter Debrass's Hawk disengaged on take-off and the aircraft fell into the sea ahead of the carrier. The ship's captain ordered full reverse so that the propellers would push the pilot away from the carrier, while Debrass waited for the carrier pass over him before he ejected at what is thought to be around 30 to 50 feet underwater, a world record. After some difficulty, he surfaced and was rescued.

The Sea Hawks were replaced by the BAE Sea Harrier and were finally withdrawn from service in 1984.

> The first completely successful British jet-powered carrier fighter.
>
> Tom Cleaver

Above: The P.1040 prototype VP401 first flew on 2 September 1947, piloted by Bill Humble. Buffeting led to a re-design of the rear jetpipe fairings and the addition of a bullet fairing on the tail. (Warbird exchange)

Left: Hawker Seahawk taking off from HMS *Eagle*, May 1952. (Gareth Bailey)

Above: Royal Navy 806 Squadron AWA Sea Hawk F1s unfold their wings as they taxi for take-off at RNAS Brawdy on 4 June 1953. (Russell Adams via Jet Age Museum)

Right: The same four Sea Hawks in their element over the sea near Trevine, Pembrokeshire. The 'Ace of Diamonds' were the first and last to fly the Sea Hawk. (Russell Adams via Jet Age Museum)

Sea Hawks from either 802 or 804 Squadron launching from the deck of HMS *Eagle* on a stormy day in mid-1955. (Russell Adams via Jet Age Museum)

Sea Hawk FGA.5 WV800, J-187, of 806 Squadron, Royal Navy, crashed on 15 August 1955 after landing on HMS *Eagle* when the arrestor hook broke off and the aircraft ran off the end of the flight deck. Pilot Lt Davey was recovered safely. (Keith Davey via Andy Davey)

The entire aircraft contingent on the deck of HMS *Eagle* on arrival at Valetta, Malta, in 1955. (Keith Davey via Andy Davey)

Sea Hawk FGA6 XE378 was allocated to 804 Squadron in December 1958 and was embarked aboard HMS *Albion* (hence the A on the tail) for a tour in the Far East from 20 October 1958 to 16 August 1959. In mid-February they were disembarked at RANAS Nowra. This aircraft was sold to the Indian Navy in November 1963 as IN181. (Kim Dunstan)

A Sea Hawk leaving the deck of HMS *Centaur* in the Gulf of Aden in 1959. Note the crew lying flat on the deck on the port side. (Owen Edwards via Rob Pennycook)

Sea Hawk FGA6 aircraft on the foredeck of HMS *Centaur*. XE458 was sold to the Indian Navy, while XE380 was scrapped in 1967. (Owen Edwards via Rob Pennycook)

A Sea Hawk mass start, the smoke coming from the explosive cartridge used to drive a small turbine engine connected by gears to the compressor shaft. (Russell Adams via Jet Age Museum)

Armstrong Whitworth Sea Hawk FGA6 XE399, c/n AW6343, on parade at the 1961 Abbotsinch open day. Delivered in June 1955, the aircraft served aboard HMS *Eagle* with NAS 899 during the Suez Crisis. It was sold for scrap in 1967. (Gordon Macadie)

Delivered from July 1957 to January 1958, the Dutch Sea Hawks were similar to the British Mk 6 but were fitted with Philips radio equipment. In the background are a mix of Grumman TBM-3S2 and 3W Avengers. (Rinus Verschuren)

The smoky cartridge start of Sea Hawks on the busy deck of the *Karel Doorman*. (Prudent Staal)

Sea Hawk F Mk 50 being prepared for take-off. (Prudent Staal)

The pilot waits for the launch signal from the aircraft director; the catapult bridle attaching the aircraft to the catapult shuttle can clearly be seen. (Prudent Staal)

Ground crew releasing the arrestor hook after a successful landing. (Prudent Staal)

The Royal Navy Historic Flight's Sea Hawk, WV908, was delivered in 1955 as an FGA4. After conversion to FGA6 standard the aircraft served until 1962, when it was put in store, and then in 1971 was sent to RAF Halton as a systems trainer. It was restored to flying condition at RNAS Culdrose in 1977/78 and then transferred to RNAS Yeovilton in September 1982. This is the only airworthy Sea Hawk in the world. (Richard Enfield)

Hawker Sea Hawk FGA6 WV832, c/n 6077, served with the FAA for ten years and was sold to the Indian Navy in 1964. (Gordon Macadie)

Hawker (Armstrong Whitworth) P-1040 Sea Hawk FGA Mk 6 of Squadron 300, the famous 'White Tigers' of the Indian Navy, prior to 6 July 1960 at Royal Naval Air Station Brawdy. (Steve Griff)

Supermarine

Attacker

Famous test pilot Jeffrey Quill was at the controls of the Attacker prototype TS409 when it first flew on 27 July 1946. It was designed to meet a Ministry of Supply specification for a single-engine jet fighter powered by the Rolls-Royce RB41 Nene centrifugal-flow turbojet, and used the laminar-flow wing developed for the ultimate Spitfire development, the Spiteful.

A direct result of its Spitfire heritage was the decision to adopt a tailwheel rather than a tricycle undercarriage. The jet exhaust had to be directed slightly upward to prevent damage to the landing surface. After the initial flights, it was necessary to add a fillet ahead of the tailfin and airbrakes to slow down the streamlined aircraft, which did not have the drag of a propeller.

Taking technological developments from the Sea Fang, a naval version of the Spiteful, a carrier-based version of the Attacker was developed with folding wing tips, strengthened landing gear and an arresting hook. This aircraft also boasted a Martin-Baker ejection seat and provided for rocket-assisted take-off gear.

Supermarine test pilot Mike Lithgow flew the Attacker prototype to Paris to demonstrate the new fighter to the French air force and navy. He was afforded the unique opportunity to fly down the Champs Élysées at low altitude, passing over the Arc de Triomphe at 600 mph and returning to Huln in under 25 minutes.

In September 1948 the FAA ordered 181 aircraft, armed with twin Hispano Mark V 20 mm cannon in each wing and the option to fit belly fuel tanks holding 1,136.5 litres. The FB2 was fitted with a marginally improved Nene 102 engine and a framed canopy. On 22 August 1951 the Attacker entered service with the 800 Squadron FAA at Ford, West Sussex, becoming the first jet fighter to serve with the Royal Navy, landing on HMS *Eagle* on 4 March.

The sole export order was for thirty-six de-navalised Attackers for the Royal Pakistan Air Force. Fred Isaacs, who flew the Attackers in Pakistan, remembers that:

> Our check-out was conducted by Vickers Test pilot "Pee Wee" Judge. It was simply being handed a copy of Attacker "Pilots Notes" to read, being asked if there were any questions and then into the cockpit to fly. There was no special protective gear available or issued at that time; you flew with what you had.

Isaacs noted that 'one had to be physically strong to perform aerobatics in the Attacker as without power assistance the stick had to be manhandled'. The new Pakistani jets practised for the 1952 Independence Day air display, performing a 20-minute routine that ended with a spectacular vertical bomb burst split-up.

After short service in the front line with the FAA, Attackers went on to serve in trial and training units. Its low acceleration on take-off and reduced drag on landing made it an ideal aircraft for pilots to convert from piston to jet flight and to train deck officers, only being replaced when the mirror landing aid was introduced.

Supermarine Attacker F1WA485 featured in the 1952 film *The Sound Barrier* directed by David Lean. This is a fictional story about one of the problems of the new jet age: the attempt to 'break' the sound barrier that occupied the minds of designers and pilots of the time.

Mike Lithgow set a world speed record in of 909 kph (565 mph) over a 100-kilometre closed course in the Attacker on 27 February 1948. However, the Attacker did not represent a major step forward in jet technology and, although in service by 1951, giving the FAA its first experience with jet carrier operations, it was replaced within four years by the Sea Hawk.

> The fact is that little was known about the operations of jets from carriers and we were all very much feeling our way.
>
> Mike Lithgow

This is probably the third prototype Attacker, TS416, during deck landing trials aboard HMS *Illustrious* in the early 1950s. (Frank Broughton via David Page)

G-5-11 Supermarine Attacker carrying two 1,000-lb bombs and six rockets. The 1951 Farnborough report notes that this aircraft belonged to the Pakistani Navy, although no markings are visible. (Cliff Hall via Chris Hall)

Attacker F.1 WA473 was delivered to the Royal Navy in July 1951 and when withdrawn was placed on a pole at RNAS Abbotsinch. (Gordon Macadie)

An unidentified Attacker armed with twelve rockets on a cold and blustery deck. (Russell Adams via Jet Age Museum)

Pakistani Attacker R4001 at South Marston, Swindon, prior to delivery by air. A pair of Attackers would fly legs of 500 nautical miles, taking an average of a week to arrive in Karachi. (Pakistani Air Force)

PAF Attackers entered service with 11 Squadron in the summer of 1951, the first jets in the Pakistani armoury. (Pakistani Air Force)

The first of Generation Two

Wartime German research had shown that swept wings and tail surfaces delayed drag in jet aircraft, allowing them to reach greater speeds. The United States and Soviet Union quickly developed the very successful swept-wing Sabre F86 and MiG-15 fighters. In the United Kingdom the swept-wing advantage was known but its benefit over the straight laminar-flow wing was unclear. With little government support or other funding, the British industry adopted the low-cost option of adapting current models. The Hawker Sea Hawk and Supermarine Attacker surfaces were swept to produce the prototype Hawker Hunter and the less-than-successful Supermarine Swift.

Supermarine Swift

Supermarine's wartime success did not carry over to a new technological era and political demands to save costs. There were not the funds to build prototypes to allow for adequate development programmes and British companies began to fall behind their competitors.

Two privately funded Type 510 Swift prototypes were built that were basic Attackers with swept flight surfaces which significantly improved performance. However, from the outset the Swift experienced handling problems. Hawker had encountered development issues with the Hunter and the government agreement to produce the Swift was driven by a need to provide a backup in case the Hunter failed. Any naval interest faded as problem after problem caused delays. Sadly, the common engine failures suffered by pre-production aircraft were the result of an unauthorised change made only to those Avon engines built for the Swift.

Production began at South Marston with F1 WK196 first off the line, but the boosted flight controls failed to overcome the inherent instability of the aircraft. At high speed in simulated combat, pulling back on the control column brought the nose up and the aircraft flicked onto its back. A wing fence was fitted to the top of the wing and the wing leading edge was extended forward, but the problem was actually the location of the centre of gravity rather than airflow over the wing.

Although not fully combat ready, the Swift entered service with 56 Squadron in February 1954 and introduced the RAF to the Supersonic Age. For the pilots the cockpit was sophisticated compared to the Meteor and it featured power controls, airbrakes and, for the first time, an anti-g suit. In his book, *Swift Justice*, pilot Nigel Walpole explains why the Swift failed as an interceptor and was withdrawn in March 1955:

> Given its superior speed and with an element of surprise, it might have some success against first-generation fighters but not in a dog fight because of its large turning radius and loss of speed in the turn. [It was] concluded that the Swift F1 had four main shortcomings: poor operational

ceiling, poor manoeuvrability and handling at high Mach numbers, a tendency for the engine to surge at low speed and high angles of attack, and poor rearward visibility.

Two speed records and other tests proved that the aircraft performed well at low level and its broad fuselage provided additional fuel tanks, giving it a good operational range. Meteor PR9 reconnaissance aircraft were obsolete and worn out from operating at high speed and the replacement Hawker Hunters were years away from delivery. As an interim measure, a reconnaissance version of the Swift was developed when an F1 was modified with a longer nose to house camera equipment, followed by the F4 fitted with a higher tail, to compensate for the longer nose, a bubble canopy and, significantly, a dog-tooth wing providing better control. The 'Type 549 Swift Fighter-Reconnaissance Mark 5 (FR5)' with 'its robust airframe, precise control and stability at low level endeared itself to all pilots in this new role'. [Walpole]

In January 1956 it equipped 2 Squadron in Germany, becoming the first aircraft in the RAF fitted with reheat. The FR5s operated only in Germany, where the potential opposition was greatest, and by November 1957 they were being replaced by the Hunter FR10.

The FR5s were the pioneers of low-level, high-speed reconnaissance flights under enemy radar and played an important part in the development of terrain-following radar that today is essential for low-level aerial missions. On two occasions the aircraft won the annual NATO reconnaissance competitions. It was a solid, fast ride down in the treetops and, despite the unreliability of the fighter, Swifts turned out to be very strong and trustworthy.

> Now we are beginning to find that the problems associated with supersonic flight are not so great after all. The right shape, the right controls and we are there.
>
> Mike Lithgow [Lithgow, 157]

North American F-86A shown over Korea in 1951. (USAF)

VV106 was the first of the Type 510 Swift prototypes. Flown by Lt J. Elliot RN, it was the first swept-winged aircraft to land on and take off from an aircraft carrier. On 8 November 1950, twelve landings and rocket-assisted take-offs were carried out HMS *Illustrious*. VV106 is currently in storage in a good condition at the FAA Museum in Yeovilton, UK. (Brian Goodwin)

The second prototype, VV119, originally a Type 510, was displayed at Farnborough in 1950, as a Type 528 modified with a tricycle undercarriage; longer, pointed nose; kinked wing leading edge; larger fuselage diameter (to accommodate an afterburning tailpipe); greater fuel capacity; and provision for four cannon in the wings. It first flew as this type on 23 August 1950 and 'starred' as *Prometheus* in the film *The Sound Barrier*. (Cliff Hall via Chris Hall)

Supermarine 541 Swift WJ965 evolved from the Type 535 and was the second pre-production example of the 100 Swifts ordered for the RAF. This aircraft was lost on 10 November 1953 when Squadron Leader 'Ned' Lewis was conducting low-speed handling trials. Following a stall test, the aircraft spun into the ground, killing the pilot. (Joe Barr)

Built as an F1, Supermarine Swift WK198 became the prototype F4, fitted with re-heat and a variable incidence tail-plane. On 5 July 1953 Commander Mike Lithgow broke the London–Paris speed record, averaging 736.95 mph (1,076.9 kph), and on 26 September established a new world speed record of 737.7 mph (1187.21 kph) at Castel Idris, Libya. (Copyright BAE Systems courtesy of Brooklands Museum)

A colour view of WK198 in PRU blue at Farnborough in 1953. (Air Cdr J. W. Frost via Peter Arnold)

Following an engine failure it was assigned to Kirkham as an instructional airframe, but disappeared when this base was closed in 1957. In 1981 North East Aircraft Museum members found WK198 in a scrapyard, rescued it and began its restoration. It is now at Brooklands Museum. (Peter Clarke)

Supermarine Swift F4 WK275, Sheppards Stores Ltd. This is one of only four complete surviving Swifts, being lifted in March 2012 from its stone pillars, where it had sat since 1965. (Jet Art Aviation)

The very complete cockpit of WK275 as opened after forty years. (Jet Art Aviation)

The cockpit of Swift F7 XF113, the first production aircraft of fourteen F7s built. XF113 is on display at the Boscombe Down Aviation Collection. (Peter Clarke)

Gloster Javelin

> The Javelin was a radical aeroplane. She had triangular delta wings and her two Sapphire-type engines, situated at the sides of the fuselage, made her the most powerful fighter in the world. [Waterton, 164]

Work began in April 1949 on a high-performance day- and nightfighter to engage the predicted modern bombers. The competition to supply a night or all-weather interceptor fighter turned out to be between Gloster's GA5 and De Havilland's DH110 or Sea Vixen. The GA5 was selected by the RAF because, in spite of development problems and slightly inferior performance to the DH110, it could be fitted with a range of radar equipment with clearer displays that allowed for a tandem cockpit and the advantage of having both crew directly engaged in a mission.

The second prototype, WD808, was the first aircraft to be fitted with resonance meters used to examine airflow and vibrations and led to a re-design with a reduced sweep to the outer wing. Unfortunately, this aircraft and its pilot, Peter Lawrence, were lost in a flat stall where it plunged 11,000 feet in a minute with no forward movement. When the nose was raised past 45 degrees the large delta wing blocked the flow of air over the tailplane, rendering the elevators useless. 'Recovery from a stall was virtually impossible and the advice was to get the beast into a spin and try one recovery, if that failed then the advice was to eject!' [Alderson]

A production order for 200 aircraft placed in June 1952 saw the first delivery to 46 Squadron at Odiham in December 1955. These aircraft were armed with 30 mm Aden cannon, an AI-17 radar and were powered by twin Sapphire engines producing 8,000 lbs thrust each, a far cry from the Whittle's 1,600 lbs of a decade earlier. The FAW1 was not service-ready; with restrictions placed on loops and vertical flight, its combat ability was compromised. The Javelin 'was an impressive aircraft to look at and all those who heard it remember the unique sound of its two Sapphire engines'. [Alderson]

The little instruction pilots received was on the ground. It was only in June 1958 that production began on twenty-two T3 trainers.

> The Javelin was a much maligned aircraft, it did have its faults, but in the hands of a good pilot it was a very useful all weather interceptor. It could climb to 40,000 feet in about eight minutes and was capable of supersonic flight in a very shallow dive. Its top useful ceiling was around 55,000 but with a little encouragement it could make 60,000 feet. It had a much improved radar system providing the navigator with a greater initial intercept range and 'glory of glories' the pilot had a "Head Up Display". The main shortcoming was of course the Javelin's notoriously poor slow speed characteristics. [Alderson]

> I fell for her when she was no more than an inspired doodle on a bit of paper.
>
> William Arthur Waterton, Gloster chief test pilot

Javelin GA5 WD804 was the first prototype, flown for the first time by Bill Waterton on 26 November 1951. He experienced severe buffeting, which was thought to be caused by the jet efflux and the wing airflow mixing close to the tail. A fairing was constructed to move the hot exit gases further back but this increased the buffeting and destroyed the elevators. Waterton managed to land the aircraft and save vital technical data, for which he was awarded the George Cross. It was written off in June 1952 at RAF Boscombe Down. (Russell Adams via Jet Age Museum)

WD804 under test on a hard stand at Morton Valance, clearly still under construction and with a number of Meteors in the background. (Russell Adams via Jet Age Museum)

Gloster Javelin WD808 made its first public appearance at the SBAC show here at Farnborough in 1952. At this stage the 'back seater's' cockpit had a solid cover. (Cliff Hall via Chris Hall)

The third prototype, WT827, was the first to be fitted for operations, with radar and wing mounted 30 mm Aden cannon. First flying on 7 March 1953, it was used for engineering trials and was evaluated by the US Air Force, who criticised its range, which led to the development of two 250-gallon ventral or 'bosom' tanks. (Russell Adams via Jet Age Museum)

Javelin FAW1 XA550, seen flying over Cheltenham in the mid-1950s, was the first to have the new canopy and radome nose. Delivered at the end of 1954, it was assigned to RAF Weeton as a ground instructional airframe, 7484M, and later scrapped. (Russell Adams via Jet Age Museum)

Gloster Javelin FAW1 XA544 was the first production aircraft, seen on display here at Farnborough in 1954 having flown for the first time on 22 July. It was used as a test aircraft at Boscombe Down, after which it was assigned to ground training as 7558M and sold for scrap in May 1963. (Russell Adams via Jet Age Museum)

A Javelin formation with XA544, the first production Javelin, and XA546 led by three prototypes en route to Farnborough, 5 September 1954. (Russell Adams via Jet Age Museum)

Javelin FAW.1 XA562 was delivered to the RAF in 1955 but returned to Rolls Royce in 1963 to be used in the development of variable reheat jetpipe nozzles. Later that year it was sent to the Spadeadam Ranges and destroyed. (Rolls-Royce)

Javelin FAW.1 XA571 was taken into service at the end of 1955. By 1963, coded 7663M, it had been assigned to RAF Halton for ground handling instruction and was subsequently scrapped. (Ronnie Bell)

A line-up of production Javelin FAW1 aircraft at Morton Valance. XA552 was with the RAF before going to De Havilland, where it was fitted with Gyron Junior engines, and then in 1963 was sent as a target to the Spadeadam ranges. (Russell Adams via Jet Age Museum)

The next in line (above), XA553, was in squadron service for only two years before it was used for ground instruction at RAF Yatesbury as 7470M. Preserved from the early 1960s, it was sold in 1998 to Thunder City, Cape Town, where it is now mounted as a gate guardian. (Louis Vosloo)

De Havilland Sea Vixen

As early as 1946 the Admiralty had looked to expand on the success of the Fleet Air Arm Vampires and approached De Havilland to develop a twin-engine, all-weather carrier fighter. The design of the DH110 continued with the familiar twin tailboom layout of the Vampire and Venom, but instead of the traditional side-by-side seating in a single cockpit the observer/navigator was enclosed in the fuselage below and to the right of the pilot who sat in a cockpit offset to port. The 'coalhole' as it became known was challenging for the occupant as it only had a limited side view but was required to provide a dark enough environment for the observer to read the radar screens.

Development of the DH110 was delayed; accordingly, the navy selected the Sea Venom to replace the Sea Hornets. The famous wartime ace John Cunningham flew the first DH110, WG236, in September 1951. This aircraft became the first two-seater, twin-engine aircraft to break the 'sound barrier' in February 1952, with John Derry and Tony Richards as pilot and observer. The second DH110 – lighter and more powerful – followed in July 1952. It was this aircraft that was due to appear at the Farnborough show but was replaced by WG236, which performed a dive from 40,000 feet and broke the 'sound barrier', followed by a low-level fly past at speed and a tight turn to port. It was the tight turn that resulted in a structural failure of the starboard wing and the disintegration of the aircraft. John Derry and Tony Richards were killed instantly in the break-up, which saw both engines propelled rocket-like through the air; one landed in the watching crowd, killing twenty-nine and injuring sixty-three.

At the same time the RAF lost interest and ordered the Javelin, but the navy still required a modern all-weather fighter and it was agreed to continue the development of the DH110 concept. Significant redesign took place, incorporating the additional strengthening and equipment required for carrier operations as well as more powerful Avon engines, improved radar and an increased fuel load. The four cannons were to be retained as a back-up to the primary armament of the Blue Jay missile, or Firestreak as it became known. The first aircraft flew in June 1955 without radar and folding wings, followed by the first carrier landing aboard HMS *Ark Royal* in April 1956.

> Because of the massive control surfaces, the aircraft designers were compelled to incorporate artificial resistance to the hydraulic controls to correspond to an increase of speed. This resistance was known as "force feel" which helped to prevent aircraft overstress. [Wrigley]

A production order was given for seventy-eight aircraft with twenty-one to be used for development and testing of the first Ferranti integrated weapons system. It was also the first British fighter not to be fitted with guns; its primary armament was four Firestreak missiles, supported by twenty-eight rockets in a ventral pack.

As with so many of the early jets, training was on the job:

> Since no two seat version was built, instruction was initially carried out in the basic first-generation simulator. This made the first flight in the actual aircraft a brand new experience. Standing next to the Sea Vixen, one felt completely dwarfed. It weighed 55 tons and had two Rolls-Royce Avon 205 jet engines that produced a total of 22,500 pounds static thrust. [Wrigley]

The Sea Vixen FAW1 was a very modern fighter that could match virtually any known enemy and there was some overseas interest. However, the 1957 Defence White Paper laid out a future without manned aircraft and buyers turned to other countries to meet their requirements.

Indecision and cancellations had resulted in significant delays into service. It was a time of fast-moving technology to keep up with developments and a time of danger and courage that demanded huge commitment and dedication.

> It looked formidable with its twin booms, high tailplane and offset cockpit.
>
> *Against All Odds*, George Wrigley, Vixen pilot

De Havilland DH110 WG240, the second Vixen prototype. Fitted with 7,500-lb RA.7 engines, the all-moving 'flying' tail, powered rudders, and cambered leading edges on the outer wings, it was an impressive performer. In September 1954, WG240 made a series of preliminary approaches, overshoots and touch-and-go landings on the deck of HMS *Albion*, successfully proving that it was suitable for carrier operations. The observer's hatch is open in this view taken at Farnborough in 1952. (Cliff Hall via Chris Hall)

De Havilland DH.110 Sea Vixen FAW1 XJ476, built in 1957 and used as a trials aircraft until it was written off at Boscombe Down in 1973. It has been with the Boscombe Down Aviation Collection from 2006. (Peter Clarke)

De Havilland DH-110 Sea Vixen FAW1 XJ488, c/n10015, spent its career engaged in trials work, such as air refuelling, missile development and carrier landings and was retired in 1972 following Martel missile trails. Moved to the fire dumb at Boscombe Down, the cockpit was sent to a scrapyard. It was rescued in 1995 by Peter Burton and moved to Robertsbridge Aviation Society, Hastings. (Rolls-Royce)

HMS *Ark Royal,* 1963 – a Sea Vixen is prepared for flight. The observer's hatch is clearly visible on the right side of the pilot's position. (Andrew Patterson)

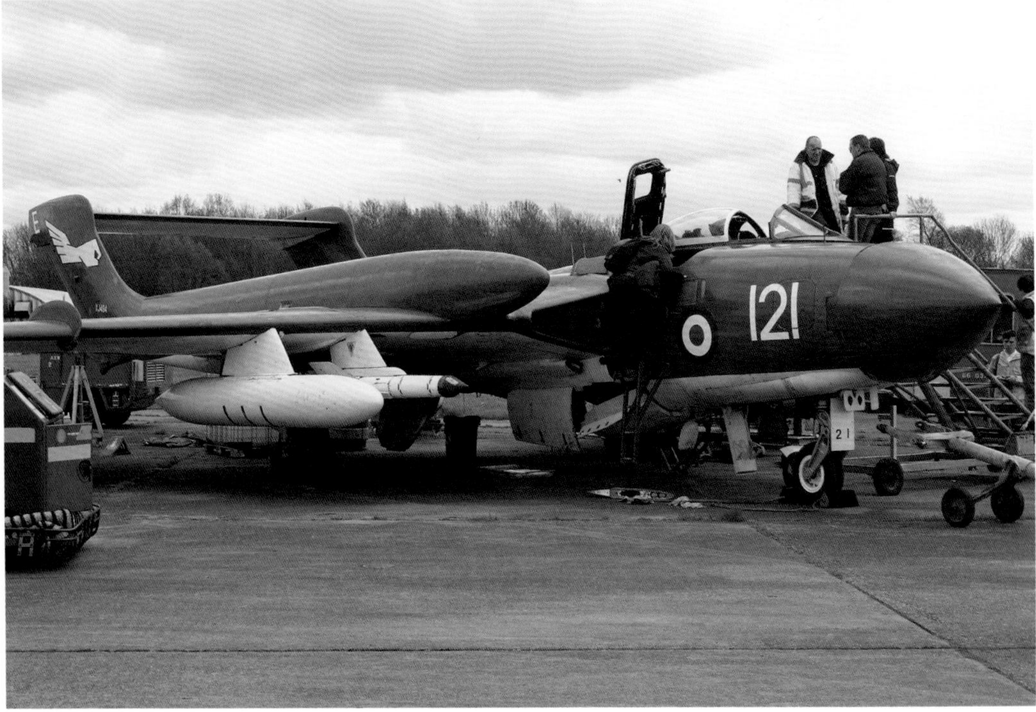

This Vixen FAW2, XJ494/E-121, is being rebuilt to taxiing status at Bruntingthorpe. (Alan Wilson)

Sea Vixen FAW 2 XP924/134/E (G-CVIX), known as *Foxy Lady*, is maintained and flown by the Fly Navy Heritage Trust. (Rob Finch)

Led by an icon of the second generation, a Hawker Hunter F4, WV325, are two De Havilland Vampire T11s, XE921 and XE893, and a Meteor T7, WH183, the difference in the generations clearly evident. (Jet Age Museum)

Bibliography

Alderson, O. W., *A Royal Air Force Pilot Remembers* (Unpublished).

Andrews, C. F. and E. B. Morgan, *Supermarine Aircraft since 1914* (London: Putnam, 1989).

Birtles, P., *De Havilland Vampire, Venom and Sea Vixen* (London: Ian Allan Ltd, 1986).

Buttler, T., 'Flying the Flat Iron', *International Air Power Review* 14 (Connecticut: AirTime Publishing, 2004).

Golley, J., *Jet: Frank Whittle and the Invention of the Jet Engine* (rev., 5th ed.) (London: Datum Publishing Limited, 2010).

Goulding, J., 'Founding the Jet Fighter Era at Glosters', *Aircraft Illustrated Annual* 1979 (London: Ian Allen Ltd, 1980).

Halley, J. J., *Broken Wings* (Tonbridge: Air Britain Historians Ltd, 1999).

Halliday, H. A., 'The Roar of the Meteor: Air Force, part 3', *Legion Magazine* (February 2010).

Harkins, H., *RAF Meteor Jet Fighters in World War II: An Operational Log* (United Kingdom: Centurion, 2013).

Higgs, G., *How Meteors Hit the Ground* (Amazon, 2013).

Horseman, M. (ed), *Aircraft Illustrated Annual 1980* (London: Ian Allen Ltd, 1979).

James, D. N., *Gloster Aircraft Company* (Stroud: Tempus, 1999).

Jones, G., *The Jet Pioneers – The Birth of Jet-Powered Flight* (London: Metheun, 1989).

Lake, J., 'Gloster Meteor Variants: Part 1', *Wings of Fame,* Vol. 14 (London: Aerospace Publishing, 1999).

Mason, F. K., *Hawker Aircraft since 1920* (London: Putnam, 1993).

Manning, C., *Fly Navy: The View From the Cockpit 1945–2000* (Barnsley: Pen & Sword Books Limited, 2000).

Nijboer, D., *Meteor I vs V1 Flying Bomb 1944* (Oxford: Osprey Publishers 2012).

Rolls-Royce, *The Jet Engine* (Derby: The Technical Publications Department, Rolls-Royce plc, 1996).

Thomas, T., *V1 Flying Bomb Aces* (Oxford: Osprey Publishers, 2013).

Walpole, N., *Swift Justice: The Supermarine Low-Level Reconnaissance Fighter* (2nd ed.) (Barnsley: Pen & Sword Books Limited, 2004).

Waterton, W. A., *The Quick and the Dead* (London: Grub Street, 2012).

Wrigley, G., *Against All Odds* (Amazon, 2014).

http://www.petermcleland.com/
http://www.thunder-and-lightnings.co.uk/
http://www.flightglobal.com/pdfarchive